How to use this book iv

UNIT 1 EXPRESSIONS AND FORMULAE

UNIT 2 RELATIONSHIPS

UNIT 3 NUMERACY

ANSWERS
www.leckieandleckie.co.uk/page/Resources/

How to use this book

Welcome to Leckie and Leckie's *National 4 Maths Practice Question Book*. This book follows the structure of the Leckie and Leckie *National 4 Maths Student Book*, so is ideal to use alongside it. Questions have been written to provide practice for topics and concepts which have been identified as challenging for many students.

Examples

Examples with worked solutions provide support for particularly tricky concepts.

Use of calculators

Questions when you should **not** use a calculator are marked with a ![icon] icon.

Questions when you could use a calculator are marked with a ![icon] icon.

Hints

Where appropriate, hints are provided to help give extra guidance and support.

Reasoning questions

Questions which require reasoning skills are marked with a ![icon] icon.

Answers

Check your own work. The answers are provided online at:

www.leckieandleckie.co.uk/page/Resources

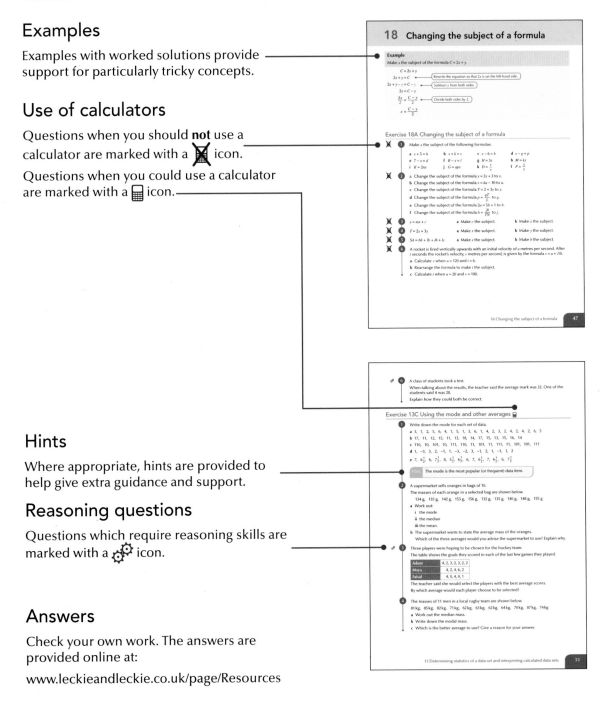

1 Using the distributive law in an expression with a numerical common factor to produce a sum of terms

Exercise 1A Using the distributive law

1 Simplify the following.

   a $3 \times k$

   c $5 \times f \times 7 \times g$

   e $(-5c) \times 6$

   g $4x \times (-8x)$

   i $5a \times (-3b) \times 2c$

   k $(-3x) \times (-9x)$

   m $\dfrac{42x}{7}$

   o $\dfrac{72x}{8y}$

   b $4 \times x \times 6$

   d $8d \times 2$

   f $7h \times 9j$

   h $\dfrac{54n}{9}$

   j $(-2m) \times (-7d)$

   l $(-4n) \times (-3d) \times 4k$

   n $\dfrac{45y}{10}$

   p $\dfrac{12a}{16b}$

2 Expand the brackets in these expressions.

   a $3(4 + m)$

   c $4(4 - y)$

   e $4(3 - 5f)$

   g $7(g + h)$

   i $6(2d - n)$

   k $5(2x - 7)$

   m $13(5 - 4y)$

   o $11(2x + 3z)$

   b $6(3 + p)$

   d $3(6 + 7k)$

   f $2(4 - 23w)$

   h $4(2k + 4m)$

   j $t(t + 5)$

   l $21(3x + 2)$

   n $12(4x - y)$

   p $9(2x - 5y + 6)$

3 Match the equivalent algebraic expressions. One has been done for you.

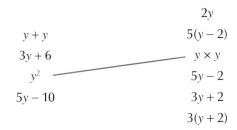

$$2y$$

$$y + y$$

$$5(y - 2)$$

$$3y + 6$$

$$y \times y$$

$$y^2$$

$$5y - 2$$

$$5y - 10$$

$$3y + 2$$

$$3(y + 2)$$

4 Expand the brackets in these expressions.

a $-3(2g - 5)$ **b** $-5(6d + 7)$ **c** $-(x - 2y)$

d $7(2x + 3y - 5)$ **e** $10(3y - 4a - 7c)$ **f** $-4(2b + 5a - 9m)$

g $-3(5c - 6g - 4k)$ **h** $-(2x + 5y)$ **i** $-4(x - 3y - 2 + 7z)$

5 Write down an expression for the area of this rectangle then simplify this expression.

$(2x - 3)$ cm

9 cm

6 Write down an expression for the area of this parallelogram then simplify this expression.

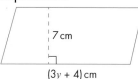

7 cm

$(3y + 4)$ cm

Hint   Area of parallelogram = base × height

7 Jeannie has written down an expression and expanded the brackets.
State where she has gone wrong, and give the correct answer.

$4(2x - 7y - 5v) = 8x - 27y + 20v$

8 Henry is x years old. Matthew is 5 years older than Henry.

a Write an expression for Matthew's age in terms of x.

b Jenny is three times older than Matthew. Write an expression for Jenny's age in terms of x. Simplify your answer.

9 Faisal has £y in his savings jar. Zain has £7 less than Faisal in his savings jar.

a Write an expression for the amount of money Zain has in his savings jar in terms of y.

b Aisha has four times as much as Zain in her savings jar. Write an expression for the amount of money in Aisha's savings jar. Simplify your answer.

1 Using the distributive law in an expression with a
numerical common factor to produce a sum of terms

2 Factorising a sum of terms with a numerical common factor

Exercise 2A Factorising

1 Find the factors of each number.

    a 24   **b** 35   **c** 37   **d** 50   **e** 54   **f** 64

2 Find the highest common factor (HCF) of each set of numbers.

    a 4 and 6   **b** 8 and 52   **c** 18 and 24

    d 28 and 42   **e** 36 and 60   **f** 12, 16 and 20

Example

Factorise $6y + 9$

$6y + 9 = 3(\ldots\ldots\ldots)$ — Find the HCF and place it outside the brackets at the front.

$= 3(2y \ldots\ldots)$ — Divide the first term by 3.

$= 3(2y + 3)$ — Divide the second term by 3.

Check: $3(2y + 3) = 6y + 9$ ✓ — Check you have the correct answer by expanding the brackets to get the expression you started with.

> **Hint** Make sure you use the **highest** common factor. Check that the terms inside the brackets in your final answer have no common factors.

3 Factorise each expression.

    a $9m + 12t$ **b** $9t + 6p$ **c** $4m + 12k$ **d** $4r + 6t$

    e $4w - 8t$ **f** $10p - 6k$ **g** $12h - 10k$ **h** $24x - 18$

    i $27y - 72$ **j** $24a + 60h$ **k** $32m + 40n$ **l** $150c + 175d$

4 Factorise each expression.

    a $4x + 2y + 6z$ **b** $20q + 15n - 45r$ **c** $12a - 36b - 54c$

    d $35k + 21j + 14$ **e** $17v - 51w + 17$ **f** $6x + 21y - 18z + 27$

5 Paul completed four questions but made errors in all of them.
In each question, find his mistake(s) and give the correct answer.

    a $8h + 12 = 2(4h + 6)$ **b** $4x - 6y = 2(2x - 6y)$

    c $10x - 25y = 5(2x + 5)$ **d** $6x - 3y + 9 = 3(2x + y + 3)$

6 Find an expression for the missing length on this rectangle.

    6 cm   area = $(18t - 12)$ cm^2

    length

3 Simplifying an expression which has more than one variable

Exercise 3A Simplifying expressions with more than one variable

1 Simplify these expressions.

   a $5t + 4t$ **b** $4m + 3m$ **c** $6y + y$ **d** $2d + 3d + 5d$

   e $7e - 5e$ **f** $6g - 3g$ **g** $3p - p$ **h** $5t - t$

   i $t^2 + 4t^2$ **j** $5y^2 - 2y^2$ **k** $4ab + 3ab$ **l** $5a^2d - 4a^2d$

2 Simplify these expressions.

   a $2x + 3y + x + 4x$ **b** $5a + 2b + 7a + 3b$

   c $7c + 3d - 2c + 4d$ **d** $8e + 5f + f - 2e$

   e $12g + 2h - 9g - h$ **f** $13i + 6j - 4i - 4j$

   g $8x + 3x^2 + 7x - 2x^2$ **h** $6m + 9n - 3m - 9n$

   i $15k + 7 - 11k - 9$ **j** $4p + 3r - 10p - r$

   k $21s - 4t - 3s + 3v - 6t - 4v$ **l** $4x - 2y - 7 - 5x - 2y - 3$

> **Hint** Remember that terms take the sign in front (to the left) of them

3 Amani completed four questions on simplifying, but made errors in all of them. In each question, find her mistake and give the correct answer.

   a $4x + 2y - x + 7y = 5x + 9y$ **b** $3x - 2x^2 - x + 7x^2 = 7x^2$

   c $6a - 2b - 4a - 8b = 2a + 6b$ **d** $5m + 3k - 4k - 9m = 4m + k$

4 Expand and simplify.

   a $3(2x + 5) + 7$ **b** $4(3y - 7) + 4$ **c** $2(5k + 8) - 11$

   d $6(2h - 11) + 3h$ **e** $4(5m - 6n) - 3m$ **f** $12(3 - a) + 6a$

5 Expand and simplify.

   a $5x + 2(x + 7)$ **b** $4y + 3(2y - 5)$ **c** $6 + 3(n + 4)$

   d $7 + 5(2b - 3)$ **e** $12 - 4(2g + 1)$ **f** $13 - (d - 6)$

6 Expand and simplify.

   a $3(2 + t) + 4(3 + t)$ **b** $6(2 + 3k) + 2(5 + 3k)$ **c** $5(2 + 4m) + 3(1 + 4m)$

   d $3(4 + y) + 5(1 + 2y)$ **e** $5(2 + 3f) + 3(6 - f)$ **f** $7(2 + 5g) + 2(3 - g)$

7 Expand and simplify.

   a $2(3 + h) - 3(5 + 3h)$ **b** $3(2g + 1) - 2(g + 5)$ **c** $2(3y + 2) - 3(3y + 1)$

   d $4(2t + 1) - 3(3t + 1)$ **e** $2(5k + 3) - 3(2k - 1)$ **f** $4(2e + 3) - 3(3e + 2)$

8 Find an expression for the perimeter of this rectangle. Give your answer in its simplest form.

$(x - 3)$ cm

$(3x + 2)$ cm

4 Evaluating an expression or a formula which has more than one variable

Example

Find the value of $5x - 2y$ when $x = 7$ and $y = 3$.

$5x - 2y = 5 \times 7 - 2 \times 3$ 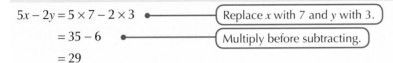 Replace x with 7 and y with 3.

$\quad\quad\quad = 35 - 6$ Multiply before subtracting.

$\quad\quad\quad = 29$

Exercise 4A Evaluating simple expressions with more than one variable

1. Find the value of $4x + 3$ when: **a** $x = 3$ **b** $x = 6$ **c** $x = 11$

2. Find the value of $3k - 1$ when: **a** $k = 2$ **b** $k = 5$ **c** $k = 10$

3. Find the value of $4 + t$ when: **a** $t = 5$ **b** $t = 8$ **c** $t = 15$

4. Evaluate $14 - 3f$ when: **a** $f = 4$ **b** $f = 6$ **c** $f = 10$

5. Find the value of $\dfrac{4d - 7}{2}$ when: **a** $d = 2$ **b** $d = 5$ **c** $d = 15$

6. Find the value of $5x + 2$ when: **a** $x = -2$ **b** $x = -1$ **c** $x = 21.5$

7. Evaluate $4w - 3$ when: **a** $w = -2$ **b** $w = -3$ **c** $w = 2.5$

8. Evaluate $10 - x$ when: **a** $x = -3$ **b** $x = -6$ **c** $x = 4.6$

9. Find the value of $5t - 1$ when: **a** $t = 2.4$ **b** $t = -2.6$ **c** $t = 0.05$

10. Evaluate $11 - 3t$ when: **a** $t = 2.5$ **b** $t = -2.8$ **c** $t = 0.99$

11. Find the value of $3x + 2y$ when:
    a $x = 7, y = 2$ **b** $x = 5, y = 4$ **c** $x = -4, y = 0.5$

12. Find the value of $4a - 3b$ when:
    a $a = 5, b = 3$ **b** $a = 2, b = 9$ **c** $a = 0.5, b = -3$

13. Find the value of $7c + 3d - 2e$ when:
    a $c = 2, d = 5, e = 3$ **b** $c = 3, d = -8, e = 0.5$ **c** $c = 5, d = 1.5, e = -6$

14. Find the value of $\dfrac{4x + 5y}{2}$ when:
    a $x = 6, y = 4$ **b** $x = 3, y = 8$ **c** $x = -2, y = 7$

15. Work out the value of each expression when $x = 17.4$, $y = 28.2$ and $z = 0.6$.
    a $x + \dfrac{y}{z}$ **b** $\dfrac{x + y}{z}$ **c** $\dfrac{x}{z} + y$

In this algebraic magic square, every row, column and diagonal should add up and simplify to $9a + 6b + 3c$.

$3a - 3b + 4c$	$2a + 8b + c$	$4a + b - 2c$
i	$3a + 2b + c$	$2a - 2b + 7c$
$2a + 3b + 4c$	**ii**	$3a + 7b - 2c$

a Copy and complete the magic square.

b Calculate the value of the 'magic number' when $a = 2$, $b = 3$ and $c = 4$.

Exercise 4B Evaluating more complex expressions, including those with more than one variable

  1 $D = 5x - y$. Find the value of D when:

    a $x = 4$ and $y = 3$ **b** $x = 5$ and $y = -3$

2 $T = y(2x + 3y)$. Find the value of T when:

    a $x = 8$ and $y = 12$ **b** $x = 5$ and $y = 7$

 3 $P = 100 - n^2$. Find the value of P when:

    a $n = 7$ **b** $n = 8$ **c** $n = 9$

4 $H = a^2 + c^2$. Find the value of H when:

    a $a = 3$ and $c = 4$ **b** $a = 5$ and $c = 12$

5 $K = m^2 - n^2$. Find the value of K when:

    a $m = 5$ and $n = 3$ **b** $m = -5$ and $n = -2$

6 $m = w(t^2 + w^2)$. Find the value of m when:

    a $t = 5$ and $w = 3$ **b** $t = 8$ and $w = 7$

7 The formula for the cost of water used by a household each quarter is:

    £32.40 + £0.003 per litre of water used.

A family uses 450 litres of water each day.

a How much is their total bill per quarter? (Take a quarter to be 91 days.)

b The family pay a direct debit of £45 per month towards their water costs.

    By how much will they be in credit or debit after the quarter?

> **Hint** 'Credit' means you have overpaid; 'debit' means you owe money.

8 A printing specialist uses a laser printer. Her profit on a print run of x pages is given by the formula $P = 4.5 + 0.02x$.

How much profit will the printing specialist make if she prints 2000 single-page race entry forms for a running club?

9 Find the value of $\sqrt{x} + 7$ when:

  a $x = 4$ **b** $x = 9$ **c** $x = 1$

10 Find the value of $x^2 - \sqrt{y}$ when:

  a $x = 3, y = 1$ **b** $x = 5, y = 4$ **c** $x = 6, y = 16$

11 Find the value of $c(\sqrt{c} + \sqrt{d})$ when:

  a $c = 9, d = 4$ **b** $c = 16, d = 1$ **c** $c = 4 , b = 25$

12 Find the value of $\sqrt{a}(2a + b^2)$ when:

  a $a = 9, b = 3$ **b** $a = 4, b = 6$ **c** $a = 25, b = 2$

13 The area, A, of a rhombus is given by the formula

$$A = \frac{1}{2}d_1d_2$$

where d_1 and d_2 represent the two diagonals of the rhombus.

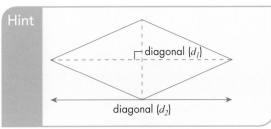

Hint

Find the area of a rhombus with diagonals of length 16 cm and 10 cm.

14 Acceleration is measured using the formula

$$a = \frac{v - u}{t}$$

where a is the acceleration (in m/s²), v is the new speed (in m/s), u is the initial speed (in m/s) and t is the time taken (in seconds) to accelerate.

Find the acceleration when u is 45 m/s, v is 80 m/s and t is 5 seconds.

15 Freda makes greetings cards and calendars to sell at a Christmas fayre. She charges 80p for each card and £3 for each calendar. The money she makes from the fayre is given by the formula

$$M = 0.8g + 3c$$

where M is the money she earns, g is the number of greetings cards she sells and c is the number of calendars.

  a If Freda sells 200 cards and 35 calendars, how much money does she earn?

  b If the materials used to make the cards and calendars cost her £56, how much profit does Freda make?

16 The surface area, SA, of a right-angled triangular prism is given by the formula

$$SA = bh + lb + lh + ls$$

where b is the breadth, l is the length, h is the height and s is the side of the triangle.

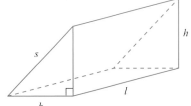

Find the surface area of a right-angled triangular prism with length 12 cm, breadth 8 cm, height 6 cm and side 10 cm.

5 Extending a straightforward number or diagrammatical pattern and determining its formula

Exercise 5A Finding the nth term of a linear sequence

1 Write down the next three terms in each sequence and describe the pattern.

    a 4, 6, 8, 10, …   **b** 3, 6, 9, 12, …   **c** 2, 4, 8, 16, …

    d 1, 4, 9, 16, …   **e** 300 000, 30 000, 3000, …   **f** 47, 40, 33, 26, …

2 For each number sequence, find the rule and write down the next three terms.

    a 7, 14, 28, 56, …   **b** 3, 10, 17, 24, 31, …   **c** 1, 3, 7, 15, 31, …

    d 40, 39, 37, 34, …   **e** 3, 6, 11, 18, 27, …   **f** 4, 5, 7, 10, 14, 19, …

    g 4, 6, 7, 9, 10, 12, …   **h** 5, 8, 11, 14, 17, …   **i** 11, 12, 14, 17, 21, …

    j 10, 9, 7, 4, …   **k** 4000, 2000, 1000, 500, …   **l** 10, 8.5, 7, 5.5, …

3 Write down the next two terms in each sequence and describe the pattern.

    a 1, 1, 2, 3, 5, 8, 13, 21, …   **b** 2, 3, 5, 8, 12, 17, …

Example

For the sequence 3, 7, 11, 15, …

a write down the next two terms

b find a rule for the nth term in the sequence

c find the 50th term.

a 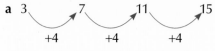   The terms in the sequence increase by 4 each time.

    $+4$ $+4$ $+4$

    $15 + 4 = 19$ $19 + 4 = 23$

    The next two terms are 19 and 23.

b

Position number	1	2	3	4	n
Position number × 4	4	8	12	16	$4n$
Term in sequence	3	7	11	15	…

Draw a table.

The sequence is formed by adding 4 each time, so multiply by 4.

nth term is $4n - 1$.

Compare 'position number × 4' with the actual term, to complete the nth term. Each term is 1 less than $4n$.

c 50th term = $4 \times 50 - 1 = 199$

Substitute 50 for n.

4 Find the nth term in each linear sequence.

    a 5, 7, 9, 11, 13 …   **b** 3, 11, 19, 27, 35, …   **c** 6, 11, 16, 21, 26, …

    d 3, 9, 15, 21, 27, …   **e** 4, 7, 10, 13, 16, …   **f** 3, 10, 17, 24, 31, …

5 For each linear sequence, **a** to **f**, find:

    i the nth term   **ii** the 50th term.

      a 3, 5, 7, 9, 11, …   **b** 5, 9, 13, 17, 21, …   **c** 8, 13, 18, 23, 28, …

      d 2, 8, 14, 20, 26, …   **e** 5, 8, 11, 14, 17, …   **f** 2, 9, 16, 23, 30, …

6 For each sequence **a** to **f**, find:

    i the nth term   **ii** the 100th term.

      a 4, 7, 10, 13, 16, …   **b** 7, 9, 11, 13, 15, …   **c** 3, 8, 13, 18, 23, …

      d 1, 5, 9, 13, 17, …   **e** 2, 10, 18, 26, …   **f** 5, 6, 7, 8, 9, …

7 The tables at a conference centre can each take three people. The tables are always put together to sit people as shown.

  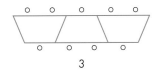

    1   2   3

  a Copy and complete the table shown.

Number of tables (n)	1	2	3	4		8
Number of people (P)	3	6				

  b Write down a formula for calculating the number of people (P) when you know the number of tables (n).

  c How many people can be seated at 12 tables?

  d How many tables does the conference centre need to set out for 50 people?

8 This pattern of shapes is built up from matchsticks.

  a Draw the fourth diagram.

  b Copy and complete the table shown.

                                  1   2   3

Diagram number (n)	1	2	3	4		9
Number of matchsticks (M)	6	11				

  c Write down a formula for calculating the number of matches (M) when you know the diagram number (n).

  d How many matchsticks are in the 25th diagram?

  e Diagram 3 is the biggest diagram that you could make with 20 matchsticks. What is the biggest diagram you could make with 200 matchsticks?

9 This pattern of hexagons is built up from matchsticks.

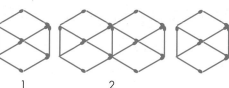

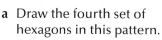

a Draw the fourth set of hexagons in this pattern.

b Copy and complete the table shown.

Diagram number (*n*)	1	2	3	4		10
Number of matchsticks (*M*)	10	19				

c Write down a formula for calculating the number of matches (*M*) when you know the diagram number (*n*).

d How many matchsticks would you need for the 60th set of hexagons?

e If you only have 100 matchsticks, which is the largest set of hexagons you could make?

10 A catering company uses tables in the shape of a trapezium. The diagrams show how many people can sit at different numbers of tables.

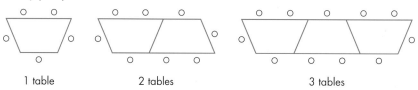

1 table   2 tables   3 tables

a Draw the next pattern.

b Copy and complete the table shown.

Number of tables (*n*)	1	2	3	4		8
Number of people (*P*)	5	8				

c Write down a formula for calculating the number of people (*P*) when you know the number of tables (*n*).

d How many people can be seated at 15 tables?

e Up to 150 people must be seated for a charity event.
How many tables arranged like this does the catering company need?

5 Extending a straightforward number or diagrammatical pattern and determining its formula

6 Calculating the gradient of a straight line from horizontal and vertical distances

Example

Find the gradients of lines A and B.

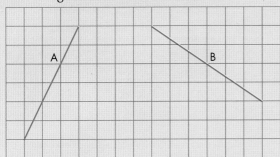

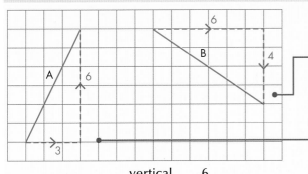

Choose any two points on the line and create a right-angled triangle. Count grid squares vertically and horizontally.

Line A: gradient = $\dfrac{\text{vertical}}{\text{horizontal}} = \dfrac{6}{3} = 2$

Line B: gradient = $\dfrac{\text{vertical}}{\text{horizontal}} = -\dfrac{4}{6} = -\dfrac{2}{3}$

The line is sloping **downwards** from left to right, so it has a **negative** gradient.

Exercise 6A Gradient of a straight line

 Work out the gradient of lines A to J.

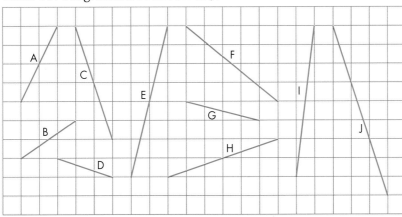

 Draw lines with these gradients.

    a 3   **b** $\frac{1}{2}$ **c** −1   **d** 8   **e** $\frac{3}{4}$ **f** $-\frac{1}{3}$

3 Work out the gradient of:

  a line A   **b** line J   **c** line D   **d** line E   **e** a line parallel to B.

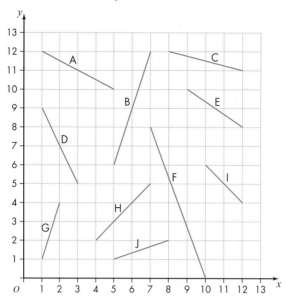

4 In the USA, the difficulty of a ski run is based on the gradient of the slope. The runs are classified as shown in the table.

Gradient of slope	Colour
less than 0.5	Green
0.5–0.8	Blue
greater than 0.8	Black

  a Calculate the gradient of the slope shown below and state its classification.

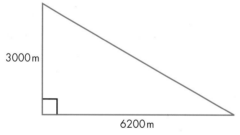

3000 m

6200 m

Hint   Convert to a decimal fraction.

  b The diagram below shows the hill that Ben skied on. Ben thinks it was a black run.

   Is Ben correct? Give a reason for your answer.

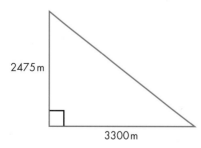

2475 m

3300 m

5 According to Health and Safety regulations, a ladder placed against a wall can have a maximum gradient of 4.

Hilary places her ladder against a wall so that the top is 8.5 m high, and the bottom is 2 m from the wall.

Does Hilary's ladder meet the regulations? Give a reason for your answer.

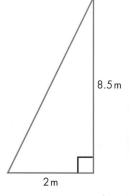

8.5 m

2 m

7 Calculating the circumference and area of a circle

Exercise 7A Circumference of a circle

For Questions 2–9, use the $\boxed{\pi}$ **button on your calculator.**

1 Copy the diagram and label it using the circle terms below.

arc   radius   tangent   diameter   sector   segment   chord   circumference

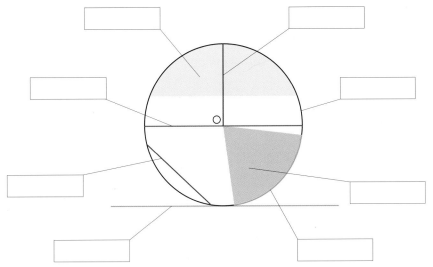

2 Calculate the circumference of each circle. Give your answers to 1 decimal place (1 d.p.).

a   b   c   d   e

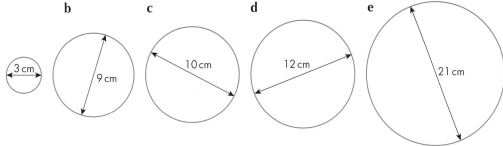

| Hint | To calculate the circumference, C, use the formula $C = \pi d$, where d is the diameter. Use the $\boxed{\pi}$ button on your calculator ($\pi \approx 3.14$). |

3 Calculate the circumference of each circle. Give your answers to 1 d.p.

a   b   c   d   e

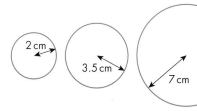

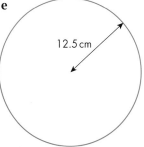

4 Pat wants to put a fence around her circular pond. The pond has a diameter of 15 m.
She plans to buy the fencing in 1-metre lengths.
How much fencing does she need?

5 Roger trains by running around a circular track that has a radius of 50 m.

a Calculate the circumference of the track. Give your answer to 1 d.p.

b How many complete circuits will he need to run to be sure of running 5000 m?

6 Calculate the perimeter of this semicircle.

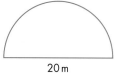

20 m

> **Hint** Divide the circumference of the circle by 2, then add the diameter.

7 What is the diameter of a circle with a circumference of 40 cm? Give your answer to 1 d.p.

8 A trundle wheel is used by surveyors to measure distances. One complete turn of the wheel is 1 m.
What is the radius of the trundle wheel?

9 The diameter of a cotton reel is 3 cm.

Cotton is wound onto the reel by rotating it on a machine.

A manufacturer wants a reel with 80 m of cotton.
How many rotations is this?

Give your answer to the nearest whole number.

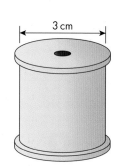

3 cm

Exercise 7B Area of a circle ▦

Use the $\boxed{\pi}$ **button on your calculator.**

1 Calculate the area of each circle. Give your answers to 1 d.p.

a   b   c   d   e

2 cm   6 cm   8 cm   10 cm   12 cm

> **Hint** To calculate the area, A, use the formula $A = \pi r^2$, where $\pi \approx 3.14$ (approx.) and r is the radius of the circle.

2 Calculate the area of each circle. Give your answers to 1 d.p.

a   b   c   d   e

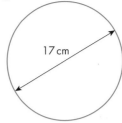

2 cm   6 cm   10 cm   17 cm   25 cm

3 The diagram shows the dimensions of Sasha's pond.

She wants to buy water lilies for the pond and would like six plants per square metre.

How many plants should she buy?

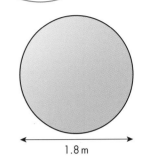

1.8 m

4 The diagram shows a circular path around a flowerbed. The radius of the flowerbed is 6 m and the width of the path is 1 m.

a Calculate the area of the flowerbed.

b Write down the radius of the large circle.

c Calculate the area of the large circle.

d Calculate the area of the path.

e Concrete costs £12 per square metre. The budget available is £300.
Is this enough for a concrete path?

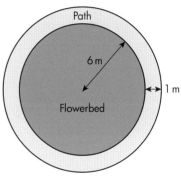

Path

6 m

1 m

Flowerbed

5 The diagram shows a metal ring.

Calculate the area of the ring. Give your answer to 1 d.p.

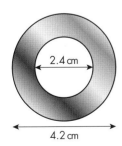

2.4 cm

4.2 cm

Hint   Calculate the area of each circle, then subtract the area of the smaller circle from the area of the larger circle.

6 The diagram shows a running track.

a Calculate the perimeter of the track. Give your answer to the nearest whole number.

b Calculate the total area inside the track. Give your answer to the nearest whole number.

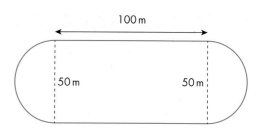

100 m

50 m   50 m

7 For each shape, calculate:

   i the perimeter   **ii** the area.

a

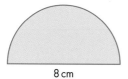

8 cm

b

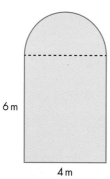

6 m

4 m

8 Helen is serving a meal for six people. Her circular table has a diameter of 80 cm.

a To sit in comfort around the table, each person needs at least 40 cm, plus 30 cm 'elbow room'.
Is the table big enough for six people to sit comfortably? Give a reason for your answer.

b Helen wants her tablecloth to overhang the table by 10 cm.
What diameter of circular tablecloth should she use?

9 A circle has a circumference of 50 cm.

a Calculate the diameter of the circle (to 1 d.p.)

b Calculate the radius of the circle (to 1 d.p.)

c Calculate the area of the circle (to 1 d.p.)

8 Calculating the area of a parallelogram, kite and trapezium

Exercise 8A Area of a parallelogram, kite and trapezium

 1 Calculate the area of each parallelogram.

a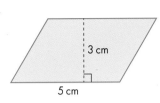
3 cm
5 cm

b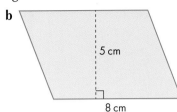
5 cm
8 cm

c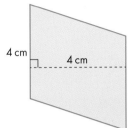
4 cm
4 cm

d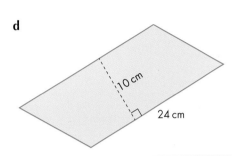
10 cm
24 cm

> **Hint** To calculate the area, A, of a parallelogram, use the formula $A = bh$, where b is the base and h is the perpendicular height of the parallelogram.

 2 Calculate the area of the shaded part of this shape.

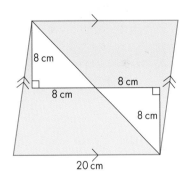
8 cm
8 cm
8 cm
8 cm
20 cm

 3 Which two of these shapes have the same area? Show your working.

A
7 cm
5 cm

B
6 cm
12 cm

C
4 cm
9 cm

 4 **a** The area of this parallelogram is 168 cm². **b** The area of this parallelogram is 275 cm².
Find the length of its base.   Find its height.

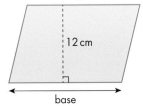
12 cm
base

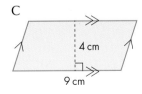
height
25 cm

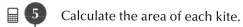

5 Calculate the area of each kite.

> **Hint** To calculate the area, A, of a kite, use the formula $A = \frac{1}{2} d_1 d_2$, where d_1 and d_2 are the diagonals of the kite.

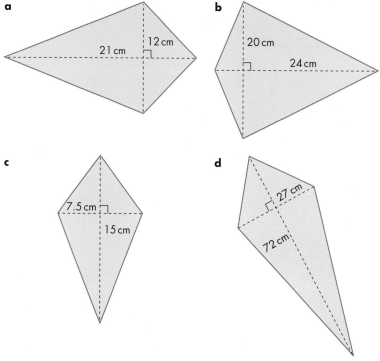

a

21 cm   12 cm

b

20 cm
24 cm

c

7.5 cm
15 cm

d

27 cm
72 cm

6 The area of this kite is 245 cm^2.

Find the length of diagonal d_1.

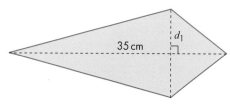

35 cm d_1

7 This shape consists of a parallelogram with a hole in the shape of a kite.

Find the shaded area of this shape.

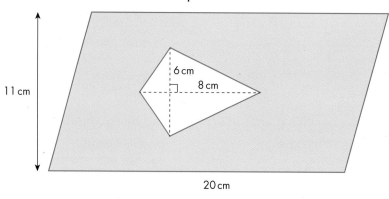

11 cm

6 cm
8 cm

20 cm

Example

Calculate the area of this trapezium.

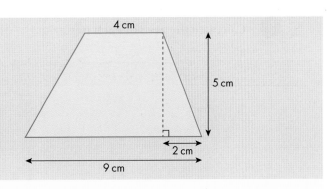

Method 1

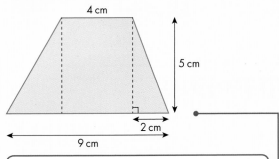

Split the shape into two triangles and a rectangle.
Work out the area of each separately.

Triangle on left: $A = \frac{1}{2}bh$

$= \frac{1}{2} \times 3 \times 5$ — Base is $9 - 2 - 4 = 3$ cm.

$= 7.5$ cm²

Triangle on right: $A = \frac{1}{2}bh$

$= \frac{1}{2} \times 2 \times 5 = 5$ cm²

Rectangle: $A = bh$

$= 4 \times 5 = 20$ cm²

Total area: $7.5 + 5 + 20 = 32.5$ cm²

Method 2

$A = \frac{1}{2}(a+b)h$ — Use the formula $A = \frac{1}{2}(a+b)h$ where a and b are the parallel sides and h is the height.

$= \frac{1}{2}(4+9) \times 5$ — Substitute $a = 4$, $b = 9$ and $h = 5$.

$= \frac{1}{2} \times 13 \times 5$

$= \frac{1}{2} \times 65$

$= 32.5$ cm²

8 Calculate the area of this trapezium.

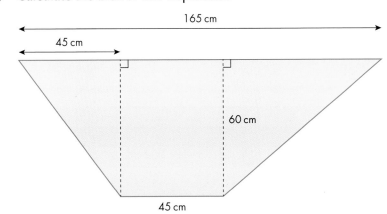

9 Calculate the area of each trapezium.

a
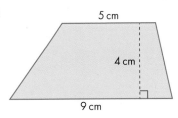

b
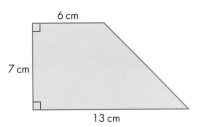

 10 Mhairi has worked out the area of this trapezium.

$A = \dfrac{1}{2}(10 + 16) \times 5$

$= (5 + 16) \times 5$

$= 21 \times 5 = 105 \, cm^2$

Mhairi has made some mistakes in her calculation.

Write out a correct solution.

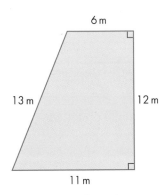

11 This is the plan of an area that is to be seeded with grass.

Seed should be planted at a rate of 30 g per m².

How much grass seed will be required?

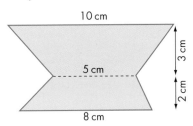

 12 A trapezium has an area of 100 cm². The parallel sides are 17 cm and 23 cm in length.

How far apart are the parallel sides?

13 Calculate the area of each of these compound shapes.

a
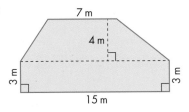

b

10 cm

5 cm

8 cm

3 cm

2 cm

14 Calculate the area of the shaded part in each diagram.

a
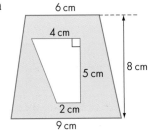

b

4 cm

3 cm

2 cm

6 cm

5 cm

9 Investigating the surface area of a prism

Example

Calculate the surface area of this cuboid.

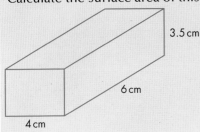

Surface area $= (2 \times 6 \times 4) + (2 \times 3.5 \times 4) + (2 \times 3.5 \times 6)$

$= 48 + 28 + 42$

$= 118\,\text{cm}^2$

Exercise 9A Surface area of prisms

 Calculate the surface area of each cuboid.

a

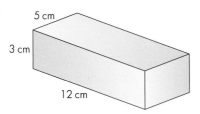

b

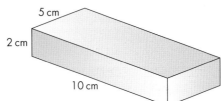

c

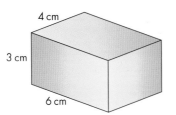

d

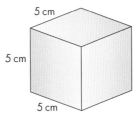

 a Calculate the surface area of each shape.

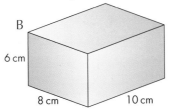

b Copy and complete these sentences.

  i The length, breadth and height of shape B is __ times as big as the length, breadth and height of shape A.

  ii The surface area of shape B is __ times as big as the surface area of shape A.

3 Calculate the surface area of this prism.

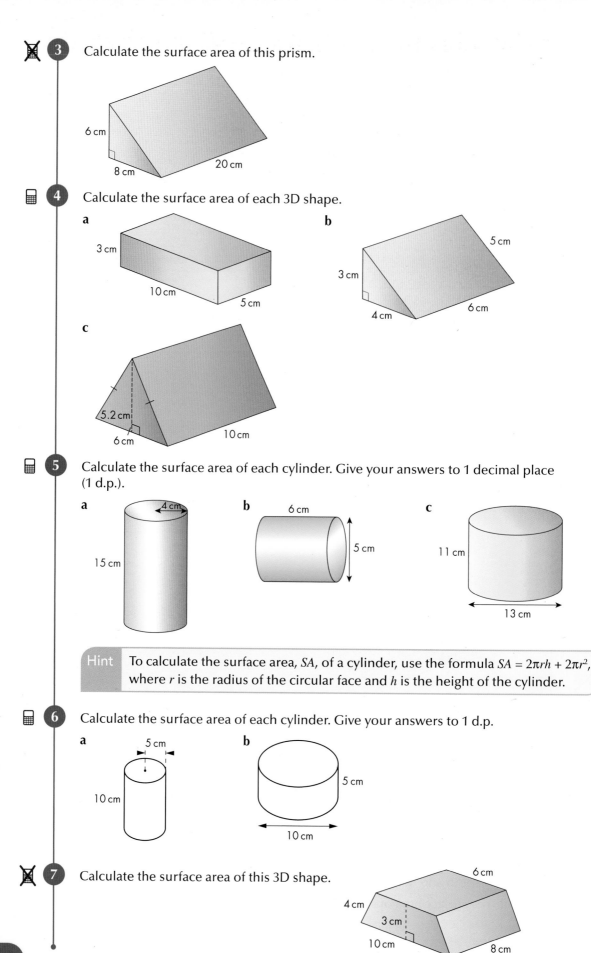

4 Calculate the surface area of each 3D shape.

a

3 cm
10 cm
5 cm

b

5 cm
3 cm
4 cm
6 cm

c

5.2 cm
6 cm
10 cm

5 Calculate the surface area of each cylinder. Give your answers to 1 decimal place (1 d.p.).

a

4 cm
15 cm

b

6 cm
5 cm

c

11 cm
13 cm

Hint   To calculate the surface area, *SA*, of a cylinder, use the formula $SA = 2\pi rh + 2\pi r^2$, where *r* is the radius of the circular face and *h* is the height of the cylinder.

6 Calculate the surface area of each cylinder. Give your answers to 1 d.p.

a

5 cm
10 cm

b

5 cm
10 cm

7 Calculate the surface area of this 3D shape.

6 cm
4 cm
3 cm
10 cm
8 cm

10 Calculating the volume of a prism

Example

Calculate the volume of this cuboid.

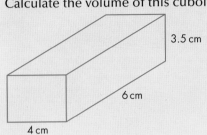

3.5 cm

6 cm

4 cm

Volume = length × breadth × height

$\quad = 4 \times 6 \times 3.5$

$\quad = 84 \text{ cm}^3$

Exercise 10A Volume of prisms

1 Calculate the volume of each cuboid.

a

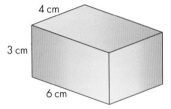

4 cm

3 cm

6 cm

b

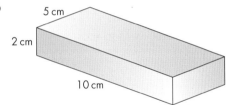

5 cm

2 cm

10 cm

c

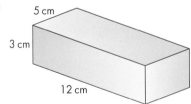

5 cm

3 cm

12 cm

d

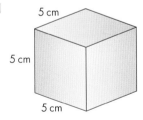

5 cm

5 cm

5 cm

2 Calculate the capacity of a swimming pool with length 12 m, breadth 5 m and depth 1.5 m.

3 Safety regulations state that, in a room where people are sleeping, there should be a minimum volume of 18 m^3 for each person.

A dormitory is 15 m long, 12 m wide and 3.5 m high.

What is the maximum number of people who can safely sleep in this dormitory?

4 Copy and complete the table for these cuboids.

	Length	Breadth	Height	Volume
a	4 cm	3 cm	2 cm	
b		3 cm	3 cm	45 cm³
c	8 cm		4 cm	160 cm³
d	6 cm	6 cm		216 cm³

 5 Calculate the volume of this prism.

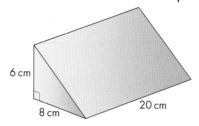

 6 Calculate the volume of each prism.

a

b

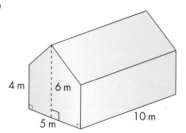

 7 Calculate the mass of each prism by finding the volume and using the density given.

a

1 cm³ has a mass of 3.13 g

b

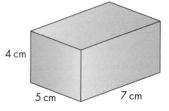

1 cm³ has a mass of 1.35 g

8 Using the density given for each shape, work out which shape is:

a the heaviest

b the lightest.

density = 1.32 g/cm³

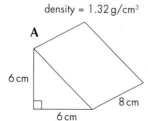
A

density = 3.13 g/cm³

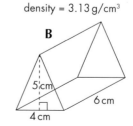
B

density = 1.35 g/cm³

C

Example

Calculate the volume of a cylinder with a radius of 4 cm and a height of 10 cm. Give your answer to 1 decimal place (1 d.p.).

Volume of cylinder = $\pi r^2 h$

$$= \pi \times 4^2 \times 10 = 502.7 \text{ cm}^3 \text{ (1 d.p.)}$$

9 Calculate the volume of each cylinder.

a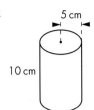
5 cm
10 cm

b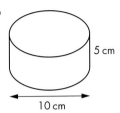
5 cm
10 cm

10 Calculate the volume of each cylinder. Give your answers to 1 d.p.

a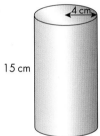
4 cm
15 cm

b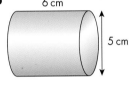
6 cm
5 cm

c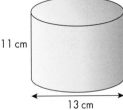
11 cm
13 cm

11 A solid iron bar is 40 cm long and has a radius of 2 cm. 1 cm³ of iron weighs 8 g.

Work out the mass of the iron bar in kilograms.

12 Andrea's mum is hosting a party and is working out how much juice to order.

Her glasses are cylindrical in shape. They have radius 1.5 cm and height 5 cm.

The juice cartons are in the shape of a cuboid that measures 30 cm by 15 cm by 18 cm.

a Calculate the amount of juice in a full carton.

b Calculate the capacity of one glass.

c How many glasses can be half filled from a carton of juice?

d There are going to be 30 guests at the party. If Andrea's mum serves each guest three half-filled glasses of juice, how many cartons of juice should she order?

11 Using rotational symmetry

Example

Rotate this shape a half-turn around the point marked with a dot to create a shape with half-turn symmetry.

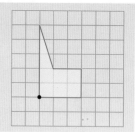

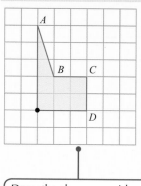

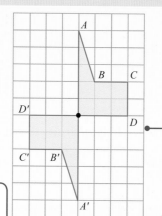

> Think about the way each point will move when it is rotated 180° (a half-turn). Label the new points *A′*, *B′*, *C′* and *D′*.

> Draw the shape on grid paper, showing the centre of rotation clearly. Working clockwise around the shape, mark each vertex with a letter, *A–D*.

Exercise 11A Rotational symmetry

1 Write down the order of rotational symmetry for each shape.

a   **b**   **c**   **d**   **e**

f   **g**   **h**   **i**   **j**

k   **l**   **m**   **n**

> **Hint** The order of rotational symmetry means the number of times a shape can be traced and rotated to fit into its own outline. You can use tracing paper to help.

2 Draw two copies of the diagram on the right.

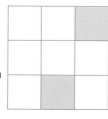

   a On the first copy, shade in two more squares so that the diagram has rotational symmetry of order 2 and no lines of symmetry.

   b On the second copy, shade in two more squares so that the diagram has rotational symmetry of order 1 and exactly 1 line of symmetry.

3 Copy each of the following shapes onto squared paper. Rotate each shape a half-turn around the point marked with a dot to create a shape with half-turn symmetry.

> **Hint** When copying each shape, allow plenty of space on your grid to draw the rotation.

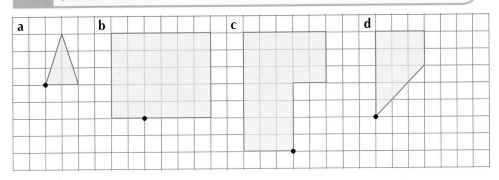

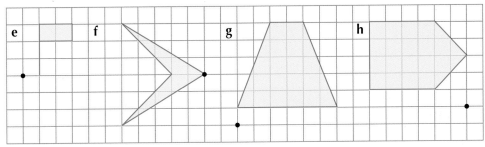

4 Copy each of the following shapes onto squared paper. Rotate each shape a quarter-turn around the point marked with a dot to create a shape with quarter-turn symmetry.

> **Hint** Remember to rotate the shape three times to create quarter turn symmetry.

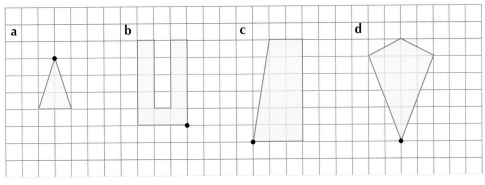

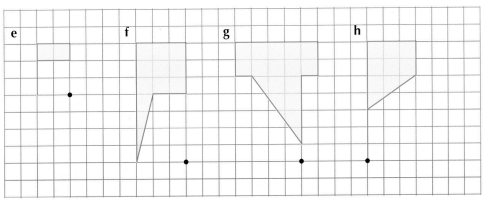

5 **a** Write down the coordinates of the labelled points on the shape.

   b Rotate the shape through a half-turn about the origin (0, 0) and write the coordinates of the rotated points.

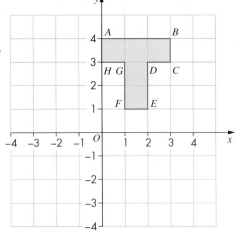

6 Rotate the shapes about centre (0, 0) to make a pattern with rotational symmetry of order 4.

a

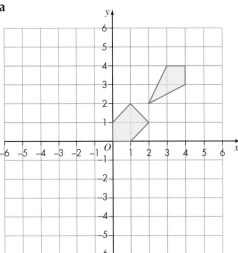

b

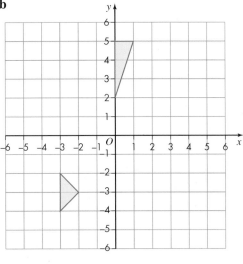

12 Constructing a frequency table with class intervals from raw data

Exercise 12A Frequency tables

1 The ages, in years, of the oldest sibling of a class of S1 pupils are shown below.

12, 13, 12, 18, 12, 15, 24, 27, 22, 23, 17, 18, 18, 20, 26, 18, 26, 18, 17, 27

Copy and complete the frequency table below.

Age (years)	Tally	Frequency
12–14		
	Total	

2 The violin grades of a group of students are shown below.

8, 3, 1, 2, 8, 3, 1, 2, 7, 8, 3, 1, 7, 6, 5, 4, 5, 5, 6, 7, 8

Draw a frequency table, with a row for each range of violin grade.

Complete your frequency table using the data given.

Violin grade	Tally	Frequency
1–2		
…		
	Total	

3 A group of pupils were asked how many times they have been abroad on holiday. Their answers were:

8, 4, 6, 2, 4, 4, 6, 9, 3, 5, 6, 3, 15, 3, 0, 11, 6, 12

Draw a frequency table, with a row for each range of number of holidays abroad.

Complete your frequency table using the data given.

Number of holidays abroad	Tally	Frequency
0–3		
…		
	Total	

4 A group of work colleagues were asked how old they were when they first rented their own home. Their ages (in years) are shown below.

31, 18, 20, 21, 27, 24, 21, 22, 23, 21, 25, 23, 25

Draw a frequency table, with a row for each range of age at first rental. The first group should be 18–20.

Complete your frequency table using the data given.

5 Class 5BD sat a maths test. Their results were:

25 18 28 21 4 11 24 16 21 23 16 21 26 22 27
12 18 23 7 29 8 13 24 17 30 19 14 18 25 20

Draw a frequency table, with a row for each range of marks. The first group should be 1–5.

Complete your frequency table using the data given.

6 The data shows the times, to the nearest minute, that 30 shoppers had to wait in the queue at a supermarket checkout.

1 3 8 12 7 4 0 9 10 15 8 1 2 7 4
2 4 7 1 0 5 4 8 4 10 7 5 4 1 5

Draw a frequency table, with a row for each range of time (in minutes). The first group should be 0–3 minutes.

Complete your frequency table using the data given.

7 A class were asked to estimate how many kilometres they had run or cycled in the last year. Their results are shown below.

34 214 653 221 536 719 527 314 213 278 56 414 242 326 436 465
791 78 658 812 15 167 285 205 512 352 418 183 192 98 14

Draw a frequency table, with a row for each range of number of kilometres. The first group should be 1–100.

Complete your frequency table using the data given.

13 Determining statistics of a data set and interpreting calculated data sets

Example

Calculate the mean of these numbers: 4, 8, 7, 5, 9, 4, 8, 3

$4 + 8 + 7 + 5 + 9 + 4 + 8 + 3 = 48$ ● ⎯⎯⎯⎯⎯⎯⎯⎯⎯ (Work out the sum of all the values.)

$$\text{Mean} = \frac{\text{sum of all values}}{\text{total number of values}}$$ ● ⎯⎯⎯⎯⎯ (There are 8 values in the data set.)

$$= \frac{48}{8} = 6$$

Exercise 13A Calculating and using the mean ▦

1 Calculate the mean of each set of data. Give your answers correct to 1 decimal place (1 d.p.) where necessary.

a 4, 2, 5, 8, 6, 4, 2, 3, 5, 1

b 21, 25, 27, 20, 23, 26, 28, 22

c 324, 423, 342, 234, 432, 243

d 17, 24, 18, 32, 16, 28, 20

e 92, 101, 98, 102, 95, 104, 99, 96, 103

f 9.8, 9.3, 10.1, 8.7, 11.8, 10.5, 8.5

2 A group of eight people took part in a marathon to raise money for charity. Their finishing times were:

2 hours 40 minutes, 3 hours 6 minutes, 2 hours 50 minutes, 3 hours 25 minutes,

4 hours 32 minutes, 3 hours 47 minutes, 2 hours 46 minutes, 3 hours 18 minutes

Calculate their mean time in hours and minutes.

3 Two families took part in a tug o' war competition.

Key family	Charlton family
Brian weighed 58 kg	David weighed 60 kg
Ann weighed 32 kg	Hannah weighed 56 kg
Steve weighed 49 kg	Pete weighed 42 kg
Alison weighed 39 kg	Barbara weighed 76 kg
Jill weighed 64 kg	Chris weighed 71 kg
Holly weighed 75 kg	Julie weighed 39 kg
Albert weighed 52 kg	George weighed 22 kg

Each family had to choose four members with a mean weight between 45 kg and 50 kg.

Choose two possible teams, one from each family.

4 The number of runs scored by a cricketer in seven innings were:
48, 32, 0, 62, 11, 21, 43

a Calculate the mean number of runs per innings.

b After eight innings, her mean score increased to 33 runs per innings.

How many runs did she score in her eighth innings?

Example

Find the median of this set of numbers: 2, 3, 5, 6, 1, 2, 3, 4, 5, 4, 6

1, 2, 2, 3, 3, 4, 4, 5, 5, 6, 6 •————————

> First write the list in numerical order.

1, 2, 2, 3, 3, ④, 4, 5, 5, 6, 6 •————————

> Find the middle number. There are 11 numbers in the list, so the middle of the list is the 6th number.

Therefore, the median is 4.

Exercise 13B Calculating and using the median

1 Write down the median of each set of data.

   a 18, 12, 15, 19, 13, 16, 10, 14, 17, 20, 11

   b 22, 28, 42, 37, 26, 51, 30, 34, 43

   c 1, −3, 0, 2, −4, 3, −1, 2, 0, 1, −2

   d 12, 4, 16, 12, 14, 8, 10, 4, 6, 14

   e 1.7, 2.1, 1.1, 2.7, 1.3, 0.9, 1.5, 1.8, 2.3, 1.4

2 The monthly wages of 11 full-time restaurant staff are:

   £820, £520, £860, £2000, £800, £1600, £760, £810, £620, £570, £650

   a Work out their median wage.

   b Calculate their mean wage.

   c How many of the staff earn more than:

      i the median wage   **ii** the mean wage?

   d Which is the better average to use? Give a reason for your answer.

3 Look at this list of numbers.

   3   3   4   7   9   10   11   14   14   15   19

   a Write down four more numbers to make the median 11.

   b Write down six more numbers to make the median 11.

   c What is the least number of values you must add to the list to make the median 3?

4 Explain why the mean may not be a good average for this data set.

   5 g   7 g   10 g   200 g   4 kg

5 The masses, in kilograms, of players in a school football team are listed below.

   68, 72, 74, 68, 71, 78, 53, 67, 72, 77, 70

   a What is the median mass of the team?

   b Calculate the mean mass of the team.

   c Which average is the better one to use? Explain why.

6 A class of students took a test.

When talking about the results, the teacher said the average mark was 32. One of the students said it was 28.

Explain how they could both be correct.

Exercise 13C Using the mode and other averages

1 Write down the mode for each set of data.

> **Hint** The mode is the most popular (or frequent) data item.

a 3, 1, 2, 5, 6, 4, 1, 5, 1, 3, 6, 1, 4, 2, 3, 2, 4, 2, 4, 2, 6, 5

b 17, 11, 12, 15, 11, 13, 18, 14, 17, 15, 13, 15, 16, 14

c 110, 10, 101, 10, 111, 110, 11, 101, 11, 111, 11, 101, 101, 111

d 1, −3, 3, 2, −1, 1, −3, −2, 3, −1, 2, 1, −1, 1, 2

e 7, $6\frac{1}{2}$, 6, $7\frac{1}{2}$, 8, $5\frac{1}{2}$, $6\frac{1}{2}$, 6, 7, $6\frac{1}{2}$, 7, $6\frac{1}{2}$, 6, $7\frac{1}{2}$

2 A supermarket sells oranges in bags of 10.

The masses of each orange in a selected bag are shown below.

  134 g, 135 g, 142 g, 153 g, 156 g, 132 g, 135 g, 140 g, 148 g, 155 g

a Work out:

  i the mode

  ii the median

  iii the mean.

b The supermarket wants to state the average mass of the oranges.

  Which of the three averages would you advise the supermarket to use? Explain why.

 3 Three players were hoping to be chosen for the hockey team.

The table shows the goals they scored in each of the last few games they played.

Adam	4, 2, 3, 2, 3, 2, 2
Maya	4, 2, 4, 6, 2
Faisal	4, 0, 4, 0, 1

The teacher said she would select the players with the best average scores.

By which average would each player choose to be selected?

4 The masses of 11 men in a local rugby team are shown below.

81 kg, 85 kg, 82 kg, 71 kg, 62 kg, 63 kg, 62 kg, 64 kg, 70 kg, 87 kg, 74 kg

a Work out the median mass.

b Write down the modal mass.

c Which is the better average to use to represent the team? Give a reason for your answer.

Example

Rachel's marks in 10 mental arithmetic tests were: 4, 4, 7, 6, 6, 5, 7, 6, 9, 6

Robert's marks in the same tests were: 6, 7, 6, 8, 5, 6, 5, 6, 5, 6

a Work out the mean and range of marks for Rachel and Robert.

b Use the mean and range to make two valid comments comparing their results.

a Rachel's mean mark: $60 \div 10 = 6$ marks

Her range: $9 - 4 = 5$ marks ●────────── ⎯⎯ Range = highest value – lowest value

Robert's mean mark: $60 \div 10 = 6$ marks

His range: $8 - 5 = 3$ marks

b On average, Rachel's and Robert's results were the same. Robert's results were more consistent.

Exercise 13D Calculating and using the range and averages 🖩

1 Work out the range for each set of data.

a 23, 18, 27, 14, 25, 19, 20, 26, 17, 24

b 92, 89, 101, 96, 100, 96, 102, 88, 99, 95

c 3.2, 4.8, 5.7, 3.1, 3.8, 4.9, 5.8, 3.5, 5.6, 3.7

d 5, −4, 0, 2, −5, −1, 4, −3, 2, 2, 0, 1, −4, 0, −2

2 The table shows the ages, in years, of a group of students on an outdoor activity course.

Age (years)	14	15	16	17	18	19
Number of students	2	3	8	5	6	1

a How many students were on the course?

b Write down the modal age of the students.

c What is the range of their ages?

3 A travel brochure shows the average monthly temperatures, in °F, for Crete.

Month	April	May	June	July	August	September	October
Temperature (°F)	68	74	78	83	82	75	72

a Calculate the mean of these temperatures.

b Write down the range of these temperatures.

c The mean temperature for Corfu was 77°F and the range was 20°F.

Compare the temperatures for Crete and Corfu.

4 The table shows the amounts taken each weekday by a sandwich shop over a 2-week period.

	Mon	Tue	Wed	Thurs	Fri
Week 1	£139	£190	£30	£219	£343
Week 2	£132	£188	£19	£203	£339

a Calculate the mean amount taken each week.

b Work out the range for each week.

c What can you say about the amounts taken for each of the 2 weeks?

13 Determining statistics of a data set and interpreting calculated data sets

14 Representing raw data in a pie chart

Example

A class of 30 pupils were asked their favourite flavour of crisps. The results are shown in the table.

Draw a pie chart to represent this information.

Flavour	Number of pupils
Salt and vinegar	13
Cheese and onion	8
Prawn cocktail	2
Ready-salted	7

Flavour	Number of pupils	Angle
Salt and vinegar	13	156°
Cheese and onion	8	96°
Prawn cocktail	2	24°
Ready-salted	7	84°
Total	**30**	**360°**

Add a third column to the table to calculate each sector angle.

Check that the angles total 360°.

Calculate the total of the frequency column.

Salt and vinegar = $\frac{13}{30} \times 360 = 156°$

Angle of sector for category = $\frac{\text{frequency of category}}{\text{total frequency}} \times 360$

Favourite flavours of crisps

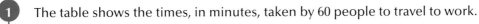

Use the calculated angles to draw the sectors of the pie chart. Use a key to show which colour represents each category.

- Salt and vinegar
- Prawn cocktail
- Cheese and onion
- Ready-salted

Exercise 14A Pie charts

1. The table shows the times, in minutes, taken by 60 people to travel to work.

Time (minutes)	10 or less	11–30	31 or more
Frequency	8	19	33

Draw a pie chart to illustrate the data.

2. The table shows the number of National 4 passes obtained by 180 pupils.

National 4 passes obtained	8 or more	6 or 7	4 or 5	3 or fewer
Frequency	20	100	50	10

Draw a pie chart to illustrate the data.

3 Tom is doing a statistics project on the use of computers and tablets. He asks 36 of his school friends about their main use of computers and tablets and records the results in the table shown.

Main use	Music/videos	Games	Homework	Social media
Frequency	5	13	3	15

a Draw a pie chart to illustrate his data.

b What conclusions can you draw from his data?

c Give reasons why Tom's data is not really suitable for his project.

4 In a survey, a TV researcher asks 120 people at a leisure centre to name their favourite type of television programme. The results are given in the table.

Type of programme	Comedy	Drama	Films	Soaps	Sport
Frequency	18	11	21	26	44

a Draw a pie chart to illustrate the data.

b Do you think the sample chosen by the researcher is representative of the population? Give a reason for your answer.

5 You are asked to draw a pie chart to represent the different breakfasts that students have one morning.

What data would you need to obtain in order to do this?

6 This table shows the favourite pets of students in Class 3B.

	Favourite pet				
	Dog	Cat	Rabbit	Guinea pig	Other
Boys	8	1	3	2	0
Girls	3	3	7	3	1

Draw pie charts to illustrate:

a the boys' favourite pets

b the girls' favourite pets

c the whole class's favourite pets.

15 Using probability

Example

A bag contains 5 red balls and 3 blue balls. A ball is taken out at random.

What is the probability that it is:

a red

b blue

c green?

a $P(\text{red}) = \frac{5}{8}$ •————[There are 5 red balls out of a total of 8.]

b $P(\text{blue}) = \frac{3}{8}$

c $P(\text{green}) = 0$ •————[There are no green balls.]

Exercise 15A Calculating probabilities

  1 Ten cards numbered from 1 to 10 (inclusive) are placed in a hat. Irene takes a card out of the hat without looking.

What is the probability that she takes out:

a the number 10

b an odd number

c a number greater than 4

d a prime number

e a number between 5 and 9

f the number 11?

  2 A bag contains 2 blue balls, 3 red balls and 4 green balls. Frank takes a ball from the bag without looking.

What is the probability that he takes out:

a a blue ball

b a red ball

c a ball that is not green

d a yellow ball

e a ball that is either red or green

f a ball that is not yellow?

  3 A bag contains 15 coloured balls. Three are red, five are blue and the rest are black. Paul takes a ball at random from the bag.

a Write down:

   i P(he takes a red)   **ii** P(he takes a blue)   **iii** P(he takes a black).

b Add together the three probabilities from part **a**. What do you notice?

c Explain your answer to part **b**.

 4 Adam, Evie, Katya, Daniel and Maria are all in the same class. Their teacher asks two of these students, at random, to tidy a cupboard.

a Write down all the possible pairings.

b How many pairs give two girls?

c What is the probability she chooses two girls?

d How many pairs give a girl and a boy?

e What is the probability she chooses a boy and a girl?

f What is the probability she chooses two boys?

5 The table shows some information about the classes in one school.

	S1		S2		S3		S4		S5	
	Boys	Girls	Boys	Girls	Boys	Girls	Boys	Girls	Boys	Girls
Pets	7	8	8	9	10	9	8	9	8	11
No pets	4	5	4	5	6	8	5	6	5	4

A representative is chosen at random from each class.

Which class has the highest probability of having a male representative with a pet?

Exercise 15B Calculating the probability of an event not happening

> **Hint** P(event not happening) = 1 − P(event happening)

1 **a** The probability of winning a prize in a tombola is $\frac{1}{25}$.
What is the probability of not winning a prize in the tombola?

  b The probability that it will rain tomorrow is 65%.
What is the probability that it will not rain tomorrow?

  c The probability that Josie wins a game of tennis is 0.8.
What is the probability that she does not win a game?

  d The probability of getting two 6s when throwing two dice is $\frac{1}{36}$.
What is the probability of not getting two 6s?

2 These letter cards are put into a bag.

A B R A C A D A B R A

  a Stan takes a letter card at random.
What is the probability that:

   i he takes a letter A   **ii** he does not take a letter A?

  b Stan takes an R from the original set of cards and keeps it. Eliza now takes a letter
from those remaining.

   i What is P(A) now?

   ii What is P(not A) now?

> **Hint** In part **b**, the total number of cards has now been reduced by 1.

3 These are the starting tiles of a board game.

START	Chelsea Park	Take a chance	London Road	Carter Road	Pay tax £500	Banner Road	Struen Road	Rest area

You throw a regular dice and move, from the start, the number of places shown by the dice.
What is the probability of *not* landing on:

  a a blue tile   **b** the Pay tax tile   **c** a coloured tile?

4 Elijah and Harris are playing a board game with a regular dice. Elijah will have to miss a turn if the next dice he rolls shows an odd number. Harris will miss a turn if the next dice he rolls shows a 1 or a 2.

Who has the better chance of *not* missing a turn the next time they roll the dice?

 5 Aaron is told that the chance he loses a game is 0.1. He says, 'So my chance of winning is 0.9.'

Give a reason why Aaron might be wrong.

Exercise 15C Calculating the probability of expected outcomes

 1 I roll an ordinary fair dice 600 times.
How many times can I expect to get a score of 1?

 2 I toss a fair coin 500 times.
How many times can I expect to get a tail?

 3 I have 5 tickets for a raffle. The probability that I win the only prize is 0.004.
How many tickets were sold altogether in the raffle?

 4 A bag contains 20 balls. 10 are red, 3 are yellow and 7 are blue. A ball is taken out at random and then replaced. This is done 200 times.
How many times would I expect to get:

a a red ball **b** a yellow or blue ball

c a ball that is not blue **d** a green ball?

 5 A headteacher is told that the probability of any student being left-handed is 0.14. She has 1200 students in her school.
Explain how she can work out how many of her students she should expect to be left-handed.

 6 An opinion poll uses a sample of 200 voters in one area. 112 of them said they would vote for Party A.

a There are a total of 50 000 voters in the area.
If they all voted, how many would you expect to vote for Party A?

b The poll is accurate to within 10%.
Can Party A be confident of winning?

 7 **a** Franz is about to take his driving test. The chance that he passes is $\frac{1}{3}$.

His sister says: 'You are sure to pass within three attempts because $3 \times \frac{1}{3} = 1$.'
Explain why his sister is wrong.

b If Franz does fail, would you expect the chance that he passes next time to increase or decrease? Explain your answer.

16 Drawing and recognising a graph of a linear equation

Example

Write down the equation of each line shown below.

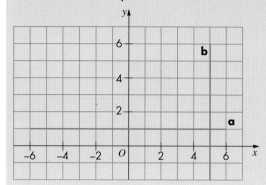

a $(-2, 1)$ $(0, 1)$ $(1, 1)$ $(4, 1)$ — Write down a few points on the line.

$y = 1$ — For each point, the y-coordinate is always 1.

b $(5, -1)$ $(5, 1)$ $(5, 2)$ $(5, 6)$ — Write down a few points on the line.

$x = 5$ — For each point, the x-coordinate is always 5.

Exercise 16A Equations of vertical and horizontal lines

 Draw a coordinate grid with x- and y-axes from −5 to 5.

a Plot these points on the grid.

$(-1, 3)$ $(4, 3)$ $(1, 3)$ $(-3, 3)$ $(0, 3)$

b Draw a line through the points you have plotted. Write the equation of the line.

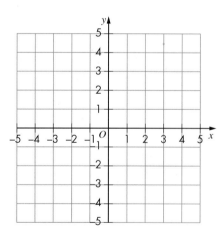

2 Draw a coordinate grid with x- and y-axes from −5 to 5.

a Plot these points on the grid.

$(4, 1)$ $(4, -2)$ $(4, 0)$ $(4, -5)$ $(4, 4)$

b Draw a line through the points you have plotted. Write the equation of the line.

 Draw a coordinate grid. Draw the graph of the lines with the following equations.

a $x = 5$ **b** $x = 2$ **c** $y = 2$ **d** $y = -3$

e $x = -2$ **f** $y = 7$ **g** $x = -4$ **h** $y = -2$

4 Write the equation of each coloured line shown.

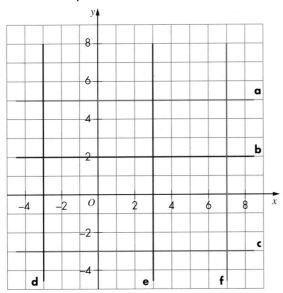

Example

x	1	2	3
y			

a Copy and complete the table of values for $y = 3x + 1$.

b Draw the line $y = 3x + 1$.

c What is the gradient of the line?

d Write the coordinates of the point where the line $y = 3x + 1$ crosses the y-axis.

a $y = 3 \times 1 + 1 = 4$ giving $(1, 4)$ — Substitute each x-coordinate into the equation.

$y = 3 \times 2 + 1 = 7$ giving $(2, 7)$

$y = 3 \times 3 + 1 = 10$ giving $(3, 10)$

x	1	2	3
y	4	7	10

b

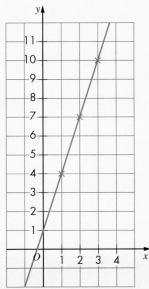

Plot the points $(1, 4)$, $(2, 7)$ and $(3, 10)$ on a coordinate grid. Connect them with a straight line.

c Gradient $= \dfrac{\text{vertical}}{\text{horizontal}}$

$= \dfrac{3}{1} = 3$

Or you can compare $y = 3x + 1$ with the equation $y = mx + c$. The value of m is the gradient.

d Line crosses y-axis at $(0, 1)$

Compare with the equation $y = mx + c$. The value of c is the y-intercept. Check by inspecting the graph you have drawn.

1 For each equation:

  i copy and complete the table of values

  ii draw the graph on a pair of coordinate axes.

  a $y = x - 4$

x	1	2	3
y			

  b $y = 2x + 1$

x	0	1	2
y			

  c $y = 4x - 2$

x	0	1	2
y			

  d $y = 5x$

x	0	1	2
y			

  e $y = -2x + 4$

x	0	1	2
y			

2 **a** Copy and complete the table of values for $y = 2x + 3$.

x	1	2	3
y			

  b Draw the graph of the line $y = 2x + 3$.

  c What is the gradient of the line?

  d Write the coordinates of the point where the line $y = 2x + 3$ crosses the y-axis.

3 **a** Copy and complete the table of values for $y = -3x - 1$.

x	1	2	3
y			

  b Draw the graph of the line $y = -3x - 1$.

  c What is the gradient of the line?

  d Write the coordinates of the point where the line $y = -3x - 1$ crosses the y-axis.

4 **a** Copy and complete the table of values for $y = \frac{1}{2}x - 2$.

x	2	4	6
y			

  b Draw the line $y = \frac{1}{2}x - 2$.

  c What is the gradient of the line?

  d Write the coordinates of the point where the line $y = \frac{1}{2}x - 2$ crosses the y-axis.

5 **a** Copy and complete the table for $y = -\frac{1}{2}x + 5$.

x	−2	0	2
y			

b Draw the graph of the line $y = -\frac{1}{2}x + 5$.

c What is the gradient of the line?

d Write the coordinates of the point where the line $y = -\frac{1}{2}x + 5$ crosses the y-axis.

6 **a** On the same set of axes, draw the graphs of $y = 3x - 1$ and $y = 2x + 3$ for $0 \leqslant x \leqslant 5$.

b At which point do the two lines intersect?

7 **a** On the same axes, draw the graphs of $y = 4x - 3$ and $y = 3x + 2$ for $0 \leqslant x \leqslant 6$.

b At which point do the two lines intersect?

8 **a** On the same axes, draw the graphs of $y = \frac{x}{2} + 1$ and $y = \frac{x}{3} + 2$ for $0 \leqslant x \leqslant 12$.

b At which point do the two lines intersect?

9 **a** On the same axes, draw the graphs of $y = 2x + 3$ and $y = 2x - 1$ for $0 \leqslant x \leqslant 4$.

b Do the graphs intersect? If not, explain why.

10 For each graph **a** to **c**:

   i find the equation of the two lines

   ii describe any symmetries that you can see.

a

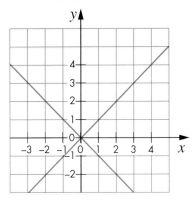

b

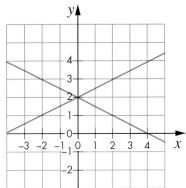

c

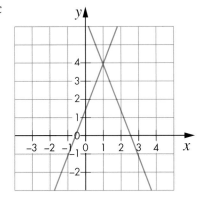

17 Solving linear equations

Example

Solve $8x - 3 = 2x + 21$.

$8x - 2x - 3 = 2x - 2x + 21$ ⟵ Subtract $2x$ from both sides of the equation to get a single term in x.

$6x - 3 = 21$

$6x - 3 + 3 = 21 + 3$ ⟵ Add 3 to both sides.

$6x = 24$

$\dfrac{6x}{6} = \dfrac{24}{6}$ ⟵ Divide both sides by 6.

$x = 4$

Check:

LHS: $8x - 3 = 8 \times 4 - 3 = 29$ ⟵ LHS = RHS so solution is correct.

RHS: $2x + 21 = 2 \times 4 + 21 = 29$

Exercise 17A Solving linear equations

1 Solve the following equations.

    a $x + 2 = 8$ **b** $y - 4 = 3$ **c** $s + 7 = 10$ **d** $t - 7 = 4$

    e $3p = 12$ **f** $5q = 15$ **g** $\dfrac{k}{2} = 4$ **h** $4n = 20$

    i $\dfrac{a}{3} = 2$ **j** $b + 1 = 2$ **k** $c - 7 = 7$ **l** $\dfrac{d}{5} = 1$

2 **a** Rafiq ran 26.9 miles less than Kathryn last week. Rafiq ran 11.1 miles.
How many miles did Kathryn run? Set up an equation to represent this information
and then solve.

    b Terry and nine of his friends went out for a meal. They split the bill equally and each
paid £10.48.
What was the total bill? Set up an equation to represent this information and then
solve.

3 Solve the following equations.

    a $2x + 5 = 13$ **b** $3x - 2 = 4$ **c** $2x - 7 = 3$ **d** $3y - 9 = 9$

    e $5a + 1 = 11$ **f** $4x + 5 = 21$ **g** $6y + 6 = 24$ **h** $5x + 4 = 9$

    i $8x - 10 = 30$ **j** $8 - x = 2$ **k** $13 - 2k = 3$ **l** $6 - 3z = 0$

4 Solve the following equations.

    a $4(x + 3) = 40$ **b** $3(x + 6) = 33$ **c** $2(2x + 7) = 22$ **d** $5(3x + 9) = 120$

    e $6(x - 3) = -12$ **f** $7(4x - 2) = -14$ **g** $4(5x + 6) = -16$ **h** $5(4x - 1) = 5$

> **Hint** Multiply out brackets, then solve as before.

5 Solve the following equations.

a $3x + 2 = x + 10$ **b** $5x + 7 = 2x + 25$ **c** $7x + 3 = 3x + 39$

d $10x + 14 = 6x + 46$ **e** $8x - 2 = 5x + 19$ **f** $7x - 6 = 2x + 9$

g $11x - 7 = 4x$ **h** $8x + 23 = 2x + 23$ **i** $9x - 15 = 6x - 3$

j $6x - 23 = 4x - 9$ **k** $5x + 14 = 8x + 23$ **l** $12x - 7 = 3x - 4$

6 Solve the following equations.

a $6(x + 2) = 4x + 22$ **b** $5(x - 1) = 2x + 16$ **c** $4(3x + 4) = 7x + 56$

d $6(3x - 5) = 5x - 4$ **e** $7(2x - 1) = 3x + 4$ **f** $3x - 50 = 8(2x - 3)$

g $5(2x + 3) = 4(x + 8) + 7$ **h** $5(1 - x) = 2(x - 8)$

7 Harry is x years old. Jemma is 5 years older than Harry.

a Write down an expression for Jemma's age in terms of x.

b Their combined age is 29.

   i Set up an equation to represent this information.

   ii Solve the equation to find the value of x.

   iii How old is Jemma?

8 The length of a rectangle is 3 cm more than its breadth.

a If x represents the breadth of the rectangle, express the length in terms of x.

> **Hint**   Sketch a rectangle and add the information given.

b The rectangle has a perimeter of 26 cm.

   i Set up an equation to represent this information.

   ii Solve the equation to find the value of x.

c What is the area of this rectangle?

Exercise 17B Solving linear inequations

1 Write down the inequality represented by each diagram.

a

b

c

d

e

f

2 Draw diagrams to illustrate these inequalities.

a $x \leqslant 2$ **b** $x > -3$ **c** $x \geqslant 1$ **d** $x < 4$

e $x \geqslant -3$ **f** $1 < x \leqslant 4$ **g** $-2 \leqslant x \leqslant 4$ **h** $-2 < x < 3$

3 Solve these inequalities and illustrate their solutions on number lines.

a $x + 5 \geqslant 9$ **b** $x + 4 < 2$ **c** $x - 2 \leqslant 3$ **d** $x - 5 > -2$

e $4x + 3 \leqslant 9$ **f** $5x - 4 \geqslant 16$ **g** $2x - 1 > 13$ **h** $3(2x + 1) < 15$

4 Solve these inequalities.

> **Hint** Solve in the same way as you solved equations with x terms on both sides of the sign. Look back at the example if you need help.

a $5x + 2 > 2x + 11$ **b** $7x - 2 < 3x + 14$

c $8x - 4 \geqslant 3x + 31$ **d** $7x + 12 \leqslant 2x + 57$

5 Mary went to the record shop with £20. She bought 2 CDs costing £x each and a DVD costing £9.50. When she got to the till, she found she didn't have enough money.

Mary took the DVD back and paid for the two CDs.

She counted her change and found she had enough money to buy a lipstick for £7.

a Explain why $2x + 9.5 > 20$ and solve the inequality.

b Explain why $2x + 7 \leqslant 20$ and solve the inequality.

c Show the solution to both of these inequalities on a number line.

d The price of a CD is a whole number of pounds.
How much is a CD?

6 Solve each linear inequality.

a $x + 3 < 8$ **b** $t - 2 > 6$ **c** $p + 3 \geqslant 11$

d $4x - 5 < 7$ **e** $3y + 4 \leqslant 22$ **f** $2t - 5 > 13$

g $2(x - 3) < 14$ **h** $3x + 8 \geqslant 11$ **i** $4t - 1 \geqslant 29$

7 Write down the values of x that satisfy:

a $x - 2 \leqslant 3$, where x is a positive integer

b $x + 3 < 5$, where x is a positive, odd integer

c $2x - 14 < 38$, where x is a square number

d $4x - 6 \leqslant 15$, where x is a positive, odd integer

e $2x + 3 < 25$, where x is a positive, prime number.

> **Hint** 'Values of x that satisfy' means the numbers which make the inequation true.

8 Frank had £6. He bought 3 cans of cola and lent his brother £3. When he reached home, he put a 50p coin in his piggy bank.
What is the most that the cans of cola could have cost?

9 The perimeter of this rectangle is greater than or equal to 10 but is less than or equal to 16.

$2x - 1$

x [rectangle]

a What are: **i** the smallest possible and **ii** the biggest possible values of x?

b What are: **i** the smallest possible and **ii** the biggest possible values of the area?

18 Changing the subject of a formula

Example

Make x the subject of the formula $C = 2x + y$.

$C = 2x + y$

$2x + y = C$ — Rewrite the equation so that $2x$ is on the left-hand side.

$2x + y - y = C - y$ — Subtract y from both sides.

$2x = C - y$

$\dfrac{2x}{2} = \dfrac{C - y}{2}$ — Divide both sides by 2.

$x = \dfrac{C - y}{2}$

Exercise 18A Changing the subject of a formula

  1 Make x the subject of the following formulae.

a $x + 5 = h$ **b** $x + k = v$ **c** $x - 6 = b$ **d** $x - g = p$

e $7 - x = d$ **f** $B - x = l$ **g** $H = 3x$ **h** $M = kx$

i $K = 2nx$ **j** $G = apx$ **k** $D = \dfrac{r}{x}$ **l** $P = \dfrac{x}{y}$

2   **a** Change the subject of the formula $y = 2x + 3$ to x.

    b Change the subject of the formula $v = ku - 10$ to u.

    c Change the subject of the formula $T = 2 + 3y$ to y.

    d Change the subject of the formula $p = \dfrac{qT}{L}$ to q.

    e Change the subject of the formula $2a = 5b + 1$ to b.

    f Change the subject of the formula $h = \dfrac{jk}{TN}$ to j.

 3 $y = mx + c$ **a** Make c the subject.   **b** Make x the subject.

4 $T = 2x + 3y$ **a** Make x the subject.   **b** Make y the subject.

 5 $SA = bh + lb + lh + ls$ **a** Make s the subject.   **b** Make b the subject.

 6 A rocket is fired vertically upwards with an initial velocity of u metres per second. After t seconds the rocket's velocity, v metres per second, is given by the formula $v = u + 10t$.

    a Calculate v when $u = 120$ and $t = 6$.

    b Rearrange the formula to make t the subject.

    c Calculate t when $u = 20$ and $v = 100$.

7 A restaurant has a large oven that can cook up to 10 chickens at a time.

The chef uses the formula:

$$T = 10n + 55$$

to calculate the length of time, T, in minutes, it takes to cook n chickens.

A large group is booked for a chicken dinner at 7 p.m. They will need a total of eight chickens.

a It takes 15 minutes to get the chickens out of the oven and prepare them for serving.

At what time should the chef put the eight chickens into the oven?

b Rearrange the formula to make n the subject

c Another large group is booked for 8 p.m. the following day. The chef calculates she will need to put the chickens in the oven at 5:50 p.m.

How many chickens is the chef cooking for this party?

8 Fern notices that the price of six coffees is 90 pence less than the price of nine teas.

Let the price of a coffee be x pence and the price of a tea be y pence.

a Express the cost of a tea, y, in terms of the price of a coffee, x.

b If the price of a coffee is £1.20, how much is a tea?

9 A rocket is fired vertically upwards with an initial velocity of u (in metres per second). After t seconds the rocket's velocity, v (in metres per second), is given by the formula $v = u + gt$, where g is a constant.

a Calculate v when $u = 120$, $g = -9.8$ and $t = 6$.

b Rearrange the formula to express t in terms of v, u, and g.

c Calculate t when $u = 100$, $g = -9.8$ and $v = 17.8$.

10 To find the perimeter of a rectangle, use the formula

$$P = 2l + 2b$$

where P is the perimeter, l is the length and b is the breadth of the rectangle.

a Find P when l is 6.2 cm and b is 4.3 cm.

b Change the subject of the formula to b.

c Using your answer to part **b**, calculate b when P is 66 cm and l is 17 cm.

11 To find the volume of a cuboid, use the formula

$$V = lbh$$

where V is the volume, l is the length, b is the breadth and h is the height of the cuboid.

a Find V when l is 5 cm , b is 4 cm and h is 3 cm.

b Change the subject of the formula to l.

c Using your answer to part **b**, calculate l when V is 300 cm³, b is 5 cm and h is 6 cm.

19 Using Pythagoras' theorem

Example

Find the length of the side marked x in each triangle. Give your answers correct to 1 decimal place (1 d.p.).

a

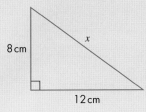

8 cm

x

12 cm

b

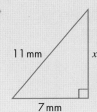

11 mm

x

7 mm

a $x^2 = 8^2 + 12^2$ — Use the formula for Pythagoras' theorem, $a^2 + b^2 = c^2$.
Identify x as the hypotenuse, so **add** the squares of the shorter sides.

$= 208$

$x = \sqrt{208}$ — Take the square root of each side.

$= 14.4$ cm — Round your answer to 1 d.p. and state the correct units.

b $x^2 = 11^2 - 7^2$ — Identify x as a shorter side, so **subtract** the squares.

$= 72$

$x = \sqrt{72}$ — Take the square root of each side.

$= 8.5$ mm — Round your answer to 1 d.p. and state the correct units.

Exercise 19A Pythagoras' theorem 🖩

1 Calculate the length of the hypotenuse of each triangle. Give your answers correct to 1 d.p.

a

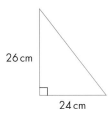

4 cm

3 cm

b

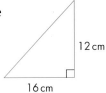

2.4 cm

3.7 cm

c

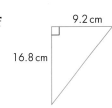

5.6 cm

9 cm

d

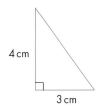

26 cm

24 cm

e

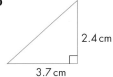

12 cm

16 cm

f

9.2 cm

16.8 cm

2 Calculate the length of x, a shorter side, for each triangle. Give your answers correct to 1 d.p.

a
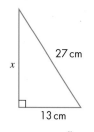
27 cm
x
13 cm

b
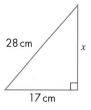
28 cm
x
17 cm

c
7.2 cm
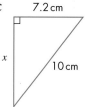
x
10 cm

d
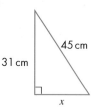
45 cm
31 cm
x

e
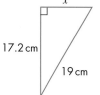
x
17.2 cm
19 cm

f
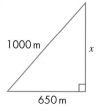
1000 m
x
650 m

g

x
2 cm
1.8 cm

h
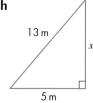
13 m
x
5 m

3 Calculate the length of the side marked x for each triangle. Give your answers to a suitable degree of accuracy.

a
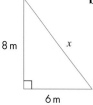
8 m
x
6 m

b
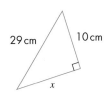
29 cm
10 cm
x

c
15 m

33 m
x

d

9.5 cm
x
8 cm

4 The diagram shows the end view of a building.
Calculate the length AB.
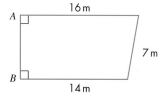
A
16 m
7 m
B
14 m

5 A pilot flies for 300 km as shown, and finds himself 200 km north of his original position.

How far east has he travelled?
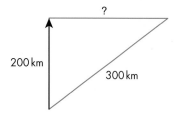
?
200 km
300 km

6 This diagram shows the cross-section of a swimming pool that is 50 m long.
The depth at the deep end is 3.5 m. The deepest part of the pool is 10 m long.

a Calculate the length of the sloping bottom of the pool, *AB*.

b The pool is 20 m wide.
What is its volume?

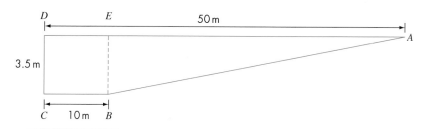

> **Hint** To calculate the volume, *V*, of a prism, use $V = Al$ where *A* is the area of the cross-section and *l* is the length of the prism.

7 A beam of wood is needed to support a sloping roof, as shown. The beam will span a horizontal distance of 3.50 m and the difference in height between the bottom and the top is 1.50 m.

A builder has a 4-metre beam.
Is this long enough?

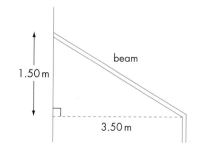

8 A ladder, 15 m long, leans against a wall.
The ladder must reach 12 m up the wall.

How far away from the wall should the foot of the ladder be placed?

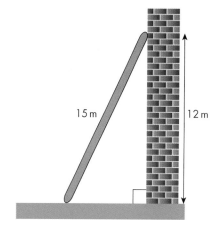

9 A rectangle is 3 m long and 1.2 m wide.
How long is the diagonal? Give your answer correct to 1 d.p.

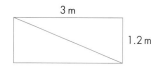

10 A ship leaves port and sails 8 km east and then sails 6 km north to a lighthouse.
What is the direct distance between the port and the lighthouse?

> **Hint** Draw a sketch to help you.

11 *A* and *B* are two points on a coordinate grid. They have coordinates (1, 3) and (2, 2).

What is the length of the straight line joining them? Give your answer to 1 d.p.

Hint　Draw a coordinate diagram to help you.

12 *A* and *B* are two points on a coordinate grid. They have coordinates (–3, –7) and (4, 6).

Show that the straight line joining them has length 14.8 units.

13 A 13-cm pencil fits exactly diagonally in a rectangular pencil tin.

The dimensions of the rectangular base are whole-number values.
What are the length and breadth of the tin?

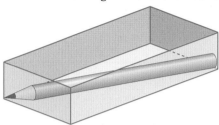

14 A rolling pin is 45 cm long.

Will it fit inside a kitchen drawer with internal measurements of 40 cm by 33 cm?
Give reasons for your answer.

15 Calculate the area of these isosceles triangles.

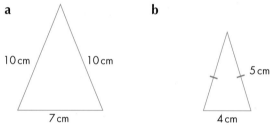

a　10 cm　10 cm　7 cm

b　5 cm　4 cm

Hint　Use Pythagoras' theorem to calculate the height of the triangle.

16 Calculate the area of an equilateral triangle of side 10 cm.

17 The diagram shows three towns, *A*, *B* and *C*, joined by two roads.
The council wants to build a road that runs directly from *A* to *C*.

How much shorter will the new road be than the two existing roads?

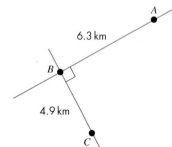

6.3 km
B
4.9 km
A
C

20 Using a fractional scale factor to enlarge or reduce a shape

Example

a Enlarge the shape shown using a scale factor of $\frac{5}{2}$

b Reduce the shape shown by a scale factor of $\frac{1}{2}$

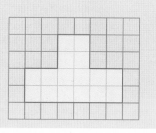

a $2 \times \frac{5}{2} = 5$ squares

$6 \times \frac{5}{2} = 15$ squares

> Multiply each side length by the scale factor, $\frac{5}{2}$, then draw the enlargement.

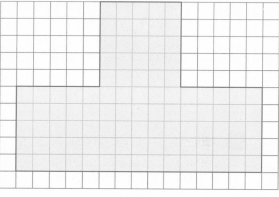

b $2 \times \frac{1}{2} = 1$ square

$6 \times \frac{1}{2} = 3$ squares

> Multiply each side length by $\frac{1}{2}$, then draw the reduction.

Exercise 20A Using fractional scale factors

Draw enlargements or reductions of the shapes in Questions 1–6 on squared paper, using the given scale factor.

a Enlarge the shape shown using a scale factor of $\frac{4}{3}$

b Enlarge the shape shown using a scale factor of $\frac{7}{3}$

c Reduce the shape shown by a scale factor of $\frac{2}{3}$

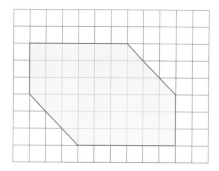

2

  a Enlarge the shape shown using a scale factor of $\frac{5}{4}$

  b Enlarge the shape shown using a scale factor of $\frac{7}{4}$

  c Reduce the shape shown by a scale factor of $\frac{3}{4}$

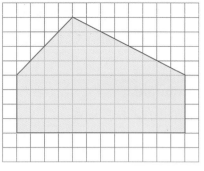

3

  a Enlarge the shape shown using a scale factor of $\frac{6}{5}$

  b Enlarge the shape shown using a scale factor of $\frac{7}{5}$

  c Reduce the shape shown by a scale factor of $\frac{3}{5}$

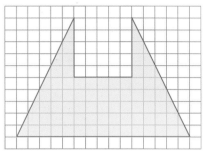

4

  a Enlarge the shape shown using a scale factor of $\frac{8}{7}$

  b Enlarge the shape shown using a scale factor of $\frac{11}{7}$

  c Reduce the shape shown by a scale factor of $\frac{6}{7}$

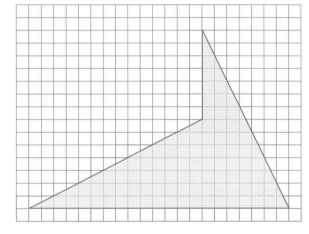

5 Using a scale factor of $\frac{6}{5}$, Pete has enlarged shape A to give shape B.

Is Pete correct? Explain your answer.

Using a scale factor of $\frac{7}{4}$, Anna has enlarged shape A to give shape B.

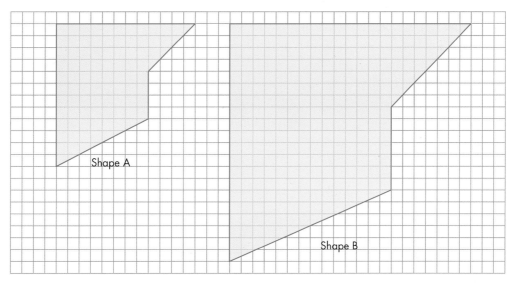

Shape A

Shape B

Is Anna correct? Explain your answer.

Exercise 20B Similar shapes

1 Which of these triangles are similar to each other?

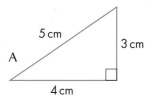

A
5 cm
3 cm
4 cm

B
33 cm
65 cm
56 cm

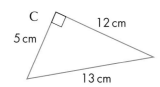

C
12 cm
5 cm
13 cm

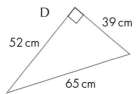

D
39 cm
52 cm
65 cm

2 Are these pairs of shapes similar? If so, give a scale factor.

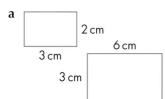

a
2 cm
3 cm
6 cm
3 cm

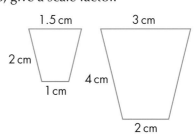

b
1.5 cm
3 cm
2 cm
1 cm
4 cm
2 cm

3 For each pair of similar shapes:

  i state the scale factor of enlargement

  ii calculate the marked x lengths.

a
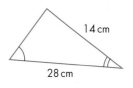

14 cm, 28 cm, x, 4 cm

b
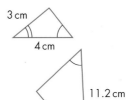

3 cm, 4 cm, 11.2 cm, x

c
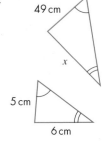
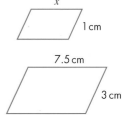
49 cm, x, 5 cm, 6 cm

d

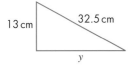
x_1, 8 cm, 9 cm, x_2, 6 cm, 9.6 cm

e

6 cm, 4 cm, 12 cm, x

f

x, 1 cm, 7.5 cm, 3 cm

4 Calculate the missing lengths in each pair of similar shapes.

a
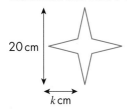
4 cm, x, 6 cm, 13 cm, 32.5 cm, y

b

13 cm, x, 22.7 cm, 21 cm

c
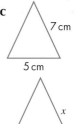
7 cm, 5 cm, x, 9 cm

5 These shapes are similar.

Calculate the value of k.

20 cm, k cm, 15 cm, 6 cm

6 In the following diagram, the triangles are similar.

Use the information shown to calculate x, the height of the statue.

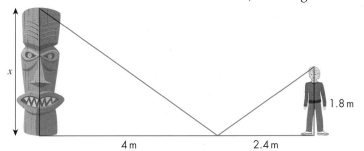

4 m   2.4 m   1.8 m

7 Ben is standing next to a tree that is 36 feet tall. The tree casts a shadow of 12 feet. Ben is 6 feet tall.

How long is Ben's shadow?

Hint   Draw a sketch to help you.

8 A 40-foot flagpole casts a shadow of 25 feet.

Work out the length of the shadow cast by a neighbouring building that is 200 feet tall.

9 A girl, 160 cm tall, stands 360 cm from a lamp post. The light from the lamp post casts a 90-cm shadow of the girl on the ground.

Work out the height of the lamp post.

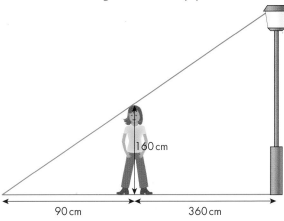

160 cm

90 cm   360 cm

Exercise 21A Calculating angles on straight and parallel lines

1 Calculate the size of the angle marked x in each diagram.

> **Hint** Remember:
> - angles on a straight line add to 180°
> - angles round a point add to 360°
> - opposite angles are equal.

a

b

c

d

e
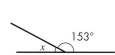

f

g
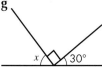

h

i

j

k

l
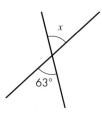

2 State the size of the lettered angles in each diagram.

> **Hint** Look for alternate (Z) angles, corresponding (F) angles and vertically opposite (X) angles.

a
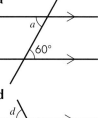

b
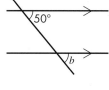

c
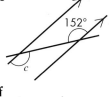

d
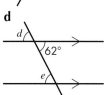

e

f

3 State the size of the lettered angles in each diagram and give a reason for your answer.

a

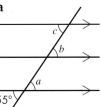

55°

b

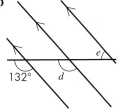

132°

c

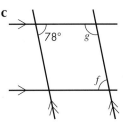

78°

Exercise 21B Calculating angles in polygons

1 Find the size of the angle marked with a letter in each diagram.

> **Hint** The angles in a triangle add to 180°.

a

50°

60°   a

b

b

60° 80°

c

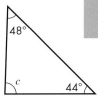

48°

c

44°

2 All the angles of a particular triangle are the same.

a What size is each angle?

b What is the name of this type of triangle?

c What is special about the sides of this triangle?

3 In the triangle on the right, two of the angles are the same.

a Work out the size of the lettered angles.

b What is the name of this type of triangle?

c What is special about the sides AB and AC of this triangle?

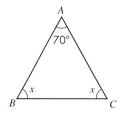

4 The diagram shows a guy rope attached to a mast on a marquee.
Find the size of the angle marked x on the diagram.

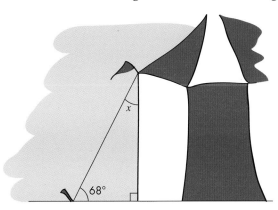

5 Find the size of the angle marked with a letter in each diagram.

a

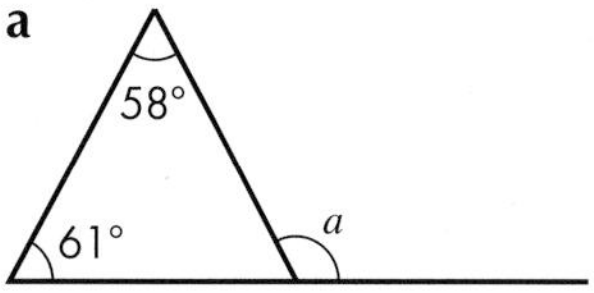

58°
61° a

b

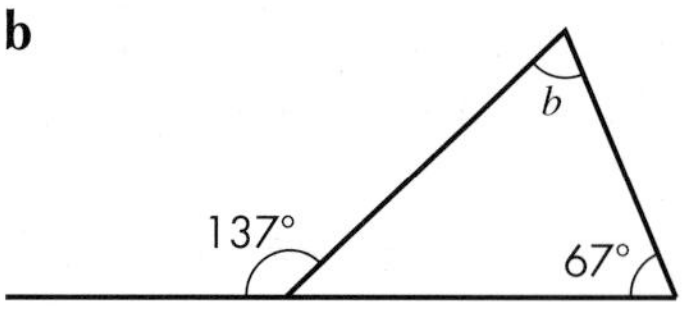

b
137° 67°

6 The diagram shows a parallelogram *ABCD*. *AC* is a diagonal.

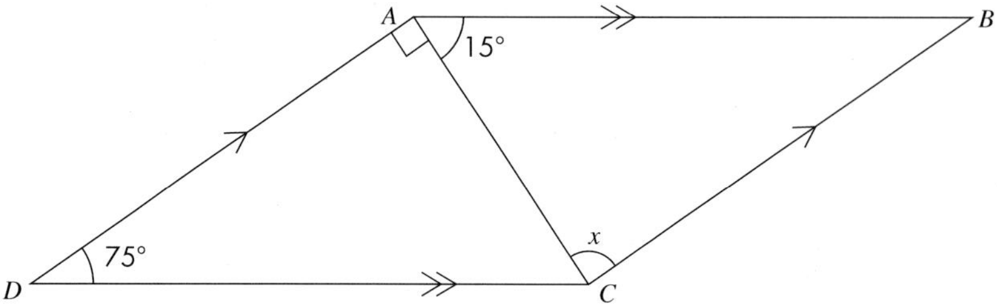

A 15° B
75°
D x C

a Show that angle *x* is 90°. **b** Calculate the size of angle *BCD*.

7 Calculate the sizes of the lettered angles in each rhombus.

> **Hint** Angles in a four-sided shape (quadrilateral) add to 360°.

a

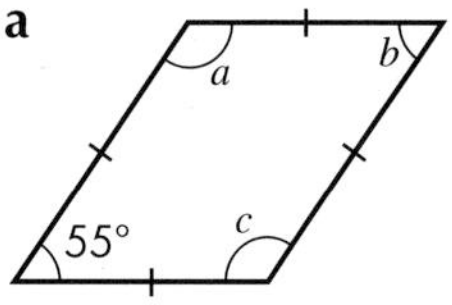

a b
55° c

b

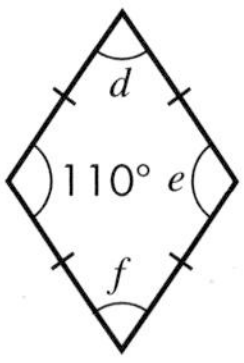

d
110° e
f

c

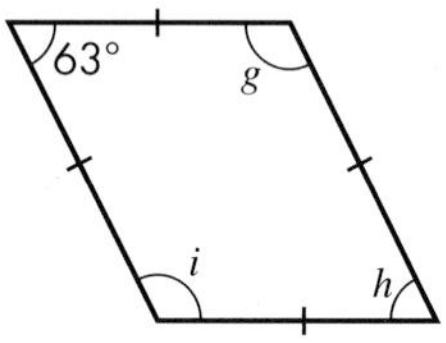

63° g
i h

8 Calculate the sizes of the lettered angles in each kite.

a

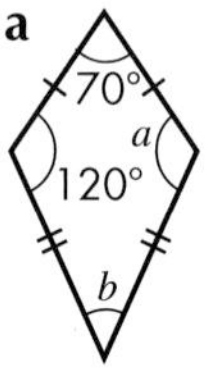

70°
a
120°
b

b

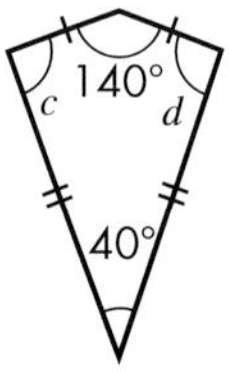

c 140°
d
40°

c

112°
96° e
f

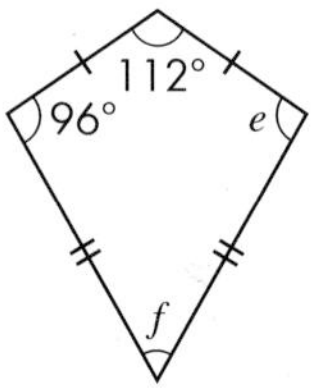

9 Calculate the sizes of the lettered angles in each parallelogram.

a

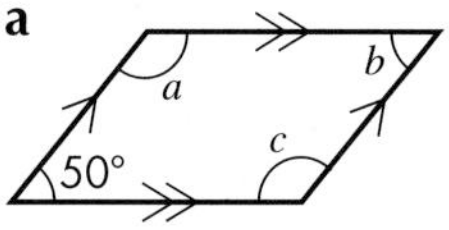

a b
50° c

b

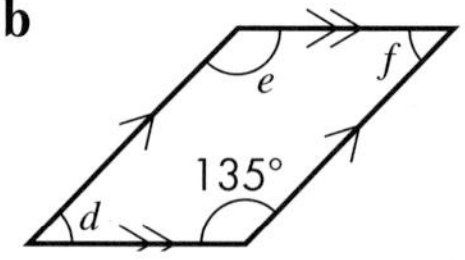

e f
135°
d

c

41° g
i
h

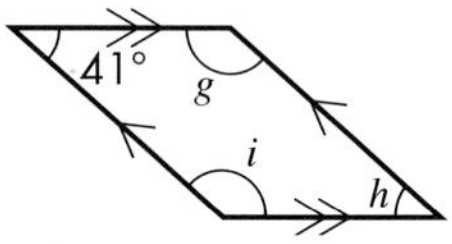

10 Calculate the sizes of the lettered angles in each trapezium.

a

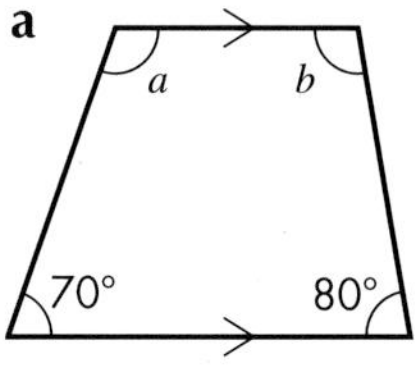

a b
70° 80°

b

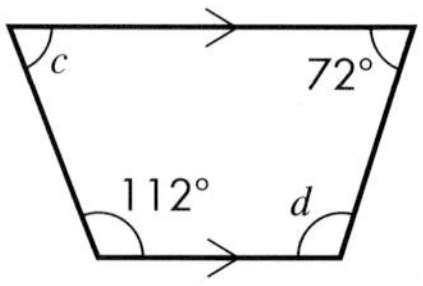

c 72°
112° d

c

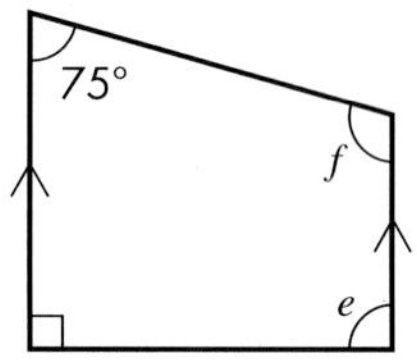

75°
f
e

11 The diagram shows the side wall of a barn.

The owner wants the angle between the roof and the horizontal to be no less than 20°, so that rain will run off quickly.

What can you say about the size of angles *BAD* and *ADC*?

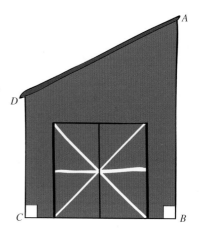

Exercise 21C Calculating angles in circles

1 Copy each diagram and fill in all the missing angles.

> Hint Two radii form an isosceles triangle.

a
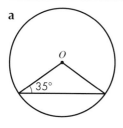

b
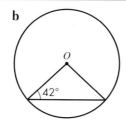

c
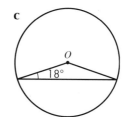

d
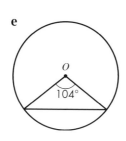

e
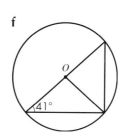

f

2 Copy each diagram, filling in all the missing angles.

> Hint The angle on the circumference of a semicircle is 90°.

a
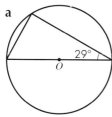

b
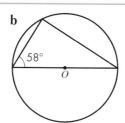

c
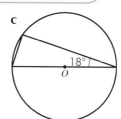

d
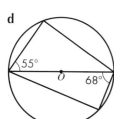

e
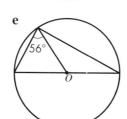

f
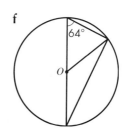

3 Copy each diagram, filling in the missing angles.

> **Hint** The tangent to a circle meets a radius at 90°.

a
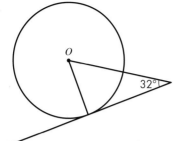

b
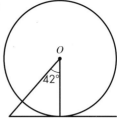

d
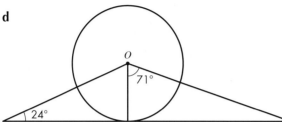

e
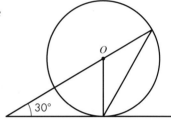

4 In the diagram, O is the centre of the circle.

Which of the following is the size of angle x marked on the diagram?

A 48°   B 57°   C 63°   D 75°

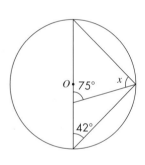

5 Each diagram shows a tangent to a circle with centre O.

Find the value of x and y in each case.

a
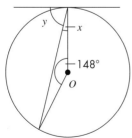

b
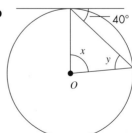

22 Calculating a side in a right-angled triangle using trigonometry

Example

Find the length of side *MP*.

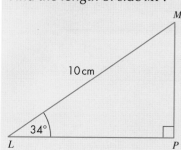

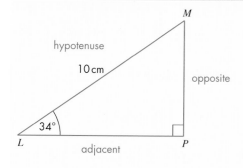

$$\sin x° = \frac{\text{opposite}}{\text{hypotenuse}}$$ • —— (Label the sides. Select the correct ratio.)

$$\sin 34° = \frac{MP}{10}$$ • —— (Substitute values for the angle, opposite and hypotenuse from the diagram.)

$$MP = 10 \times \sin 34°$$ • —— (Multiply both sides of the equation by 10.)

$$= 5.59 \text{ cm (2 d.p.)}$$

Exercise 22A Using trigonometry to calculate the length of a side 🖩

1   Use the tangent ratio to calculate the lengths of the lettered sides.
Give your answers to 2 decimal places (2 d.p.).

a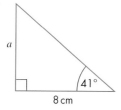

b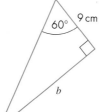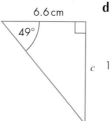

c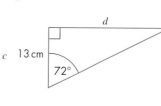

d

2 Use the sine ratio to calculate the lengths of the lettered sides.

Give your answers to 2 d.p.

a
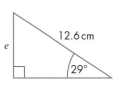

b
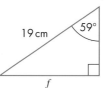

c
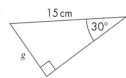

d
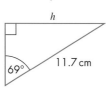

3 Use the cosine ratio to calculate the lengths of the lettered sides.

Give your answers to 2 d.p.

a

b
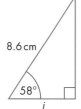

c
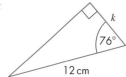

d
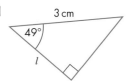

4 Calculate the lengths represented by letters. Give your answers correct to 2 d.p.

a
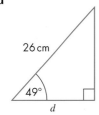

b
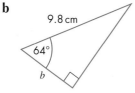

c
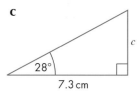

d
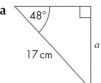

e
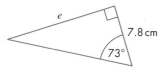

f
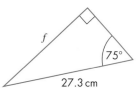

5 Calculate the height of the tree.

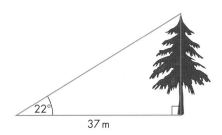

22°
37 m

6 The string of a kite makes an angle of 72° with the horizontal. The kite is 51 m vertically above a point *T*.

Calculate the horizontal distance, *y*, between the start of the string and *T*.

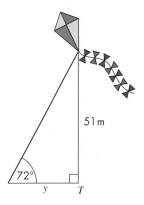

51 m

72°
y   *T*

7 A ladder, 4 m long, rests against a vertical wall. It makes an angle of 22° with the wall.

Calculate:

a how high up the wall the ladder reaches

b the distance between the foot of the ladder and the wall.

> **Hint** Draw a sketch to help you.
>
> You can solve part **b** using trigonometry, or applying Pythagoras' theorem to your answer to part **a**.

8 Angela paces out 60 m from the base of a block of flats. She then measures the angle to the top of the flats as 42°.

What is the height of the block of flats?

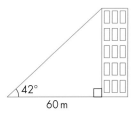

42°
60 m

9 A slide makes an angle of 46° with the ground. The slide is 7 m long.

How high above the ground is the top of the slide?

7 m
46°

23 Calculating an angle in a right-angled triangle using trigonometry

Example

Find the size of shaded angle *CHD*.

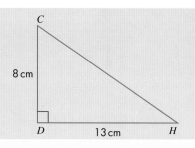

$$\tan x° = \frac{\text{opposite}}{\text{adjacent}}$$ ● — Label the sides. Select the correct ratio.

$$\tan x° = \frac{8}{13}$$ ● — Substitute values for *x*, opposite and adjacent from the diagram.

$$x° = \tan^{-1}\left(\frac{8}{13}\right)$$ ●

$$x = 31.6 \text{ (1 d.p.)}$$ — Use the inverse tan button on your calculator.

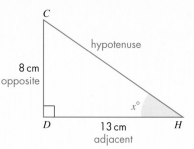

Exercise 23A Using trigonometry to calculate the size of an angle

1 Use the sine ratio to calculate the size of angle *x* in each triangle.

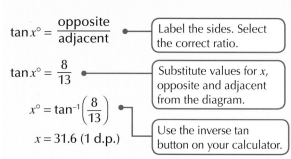

2 Use the cosine ratio to calculate the size of angle *x* in each triangle.

a
39 cm
15 cm

b
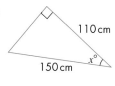
110 cm
150 cm

c
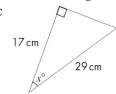
17 cm
29 cm

3 Use the tangent ratio to calculate the size of angle *x* in each triangle.

a
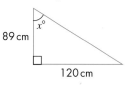
89 cm
120 cm

b

15 cm
27 cm

c

7 cm
15 cm

4 Calculate the size of the lettered angle in each triangle.

a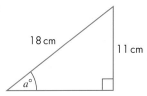

b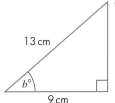

c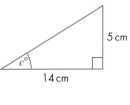

d

5 A boy was flying his kite and had let out 36 m of string when it got stuck in a tree. When the string was pulled tight, it reached the ground 27 m from the base of the tree. The tree is at right angles to the ground.

Calculate:

a the angle the string made with the ground

b how high the kite was above the ground.

> Hint   Draw a sketch to help you.

6 *ABCD* is a rectangular sheet of paper. *AC* = 21 cm and *AD* = 10 cm.

Calculate the angle *BAC*.

> Hint   A rectangle has right-angled vertices (corners).

7 Hugo is using roof trusses with the dimensions shown in this diagram.

What is the angle of slope of the roof?

> Hint   Split an isosceles triangle into two equal right-angled triangles.

8 The sensor for a security light is fixed to a house wall 2.25 m above the ground. It can detect movement on the ground up to 15 m away from the house. *B* is the furthest point where the sensor, *A*, can detect movement.

Calculate the size of angle *x*.

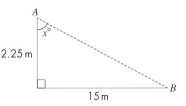

9 **a** A ladder, 8 m long, rests against a vertical wall. The foot of the ladder is 2.7 m from the base of the wall.

What angle does the ladder make with the ground?

b The ladder is positioned safely if its angle with the ground is between 70° and 80°.

Is the ladder safe? Give a reason for your answer.

24 Constructing a scattergraph and drawing and applying a best-fitting straight line

Exercise 24A Drawing and using scattergraphs

1 A newspaper advertised some used cars of the same make and model. The scattergraph shows the age and value of the cars.

  a Copy the scattergraph.

  b Draw a best-fitting line.

  > **Hint** Draw a straight line which follows the direction of the data, and has roughly the same number of points on either side of the line (ignoring any outliers).

  c Use your best-fitting line to predict the cost of a car which is $4\frac{1}{2}$ years old.

  d The following week a 7-year-old car of the same make and model was advertised in the newspaper at a cost of £9500.

  i Do you think this is a reasonable estimate? Give a reason for your answer.

  ii What other factors would affect the cost?

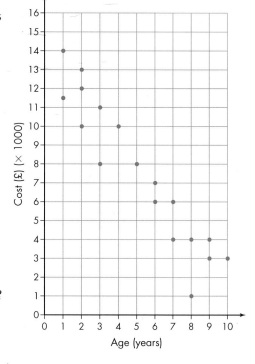

2 A teacher recorded students' attendance at maths lessons and the results in the class test at the end of the year. The results are shown in the scattergraph.

  a Copy the scattergraph.

  b Draw a best-fitting line.

  c Use your best-fitting line to predict the attendance of a student who achieved 75% in the recent test.

  d Student A does not fit the general trend. What can you say about student A?

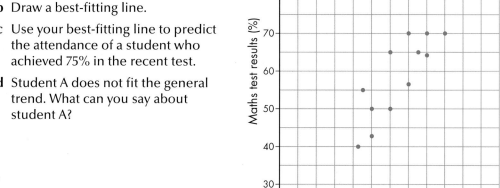

3 Aileen researched the cost of staying for one night in a 3-star hotel in Edinburgh. She noticed there was a relationship between the distance of a hotel from Waverley Station, and the cost. The table shows the results of her research.

Distance from Waverley Station (km)	0.5	1	1.5	2	2.5	2.5	3
Cost (£)	80	72	70	64	64	60	58

  a Draw a scattergraph to illustrate this information.

  b Draw a best-fitting line.

  c Use your best-fitting line to predict the cost of staying for one night in a hotel 1.75 km from Waverley Station.

  d Would this scattergraph be used to predict the cost of a hotel 20 km away? Give a reason for your answer.

4 A shopkeeper kept a record of the number of water bottles sold and the number of hours of sunshine in a day. The table shows his results.

Number of hours of sunshine	1	2	3	4	4.5	5.5	7	8
Number of water bottles sold	8	22	26	36	50	58	70	79

  a Draw a scattergraph to illustrate this information.

  b Draw a best-fitting line.

  c Use your best-fitting line to predict the number of water bottles sold on a day with 5 hours of sunshine.

5 A group of pupils sat two maths tests out of 50: one without a calculator and one with a calculator. The table shows their results.

Pupil	Anna	Beth	Craig	David	Egan	Freddy	Gary	Haia	Iain	John
Non-calculator test	13	17	22	25	20	14	14	18	12	21
Calculator test	25	30	34	13	37	29	25	44	19	35

  a Draw a scattergraph to illustrate this information.

  b Draw a best-fitting line.

  c Katie missed the Non-calculator test but scored 32 in the Calculator test. Use your best-fitting line to estimate the score she would have got on the Non-calculator test.

  d Which pupil did not do as well as expected in the Calculator test? Give a reason for your answer.

6 John sells woolly hats at a weekly market stall. He records the temperature and the number of hats sold each week. The table shows his results.

Temperature (°C)	18	14	13	12	10	6	4
Number of woolly hats sold	10	22	30	39	50	61	74

a Draw a scattergraph to illustrate this information.

b Draw a best-fitting line.

c Use your best-fitting line to predict the number of hats sold when the temperature is 10 °C.

d On a day when John sold 40 woolly hats, he estimated the temperature to be 7 °C.

Is this a reasonable estimate? Give a reason for your answer.

7 Ten people visited the doctor to have their blood pressure checked. Their ages and part of their results are shown in the table.

Age (years)	35	55	47	52	40	47	50	42	45	53
Blood pressure (mmHg)	132	170	154	161	140	155	160	145	148	167

a Draw a scattergraph to illustrate this information.

b Draw a best-fitting line.

c Use your best-fitting line to estimate the blood pressure of a 38-year-old.

8 A family were asked to measure their height and the length of their left feet. Their results are shown in the table.

Height (cm)	135	140	145	155	160	164	170	175	180
Length of left foot (cm)	22	21.6	23.7	24	26	27.5	28	29	31

a Draw a scattergraph to illustrate this information.

b Draw a best-fitting line.

c Use your best-fitting line to predict the length of the left foot of a family member with a height of 150 cm.

25 Selecting and using appropriate numerical notation and units

Example

Solve.

a $6 + 7 \times 4$

b $25 \div (12 - 7) - 2$

a $6 + 7 \times 4 = 6 + 28$ — Carry out the multiplication first.

   $= 34$

b $25 \div (12 - 7) - 2 = 25 \div 5 - 2$ — Do the brackets first.

   $= 5 - 2$ — Division is the next operation.

   $= 3$

Hint Apply the correct order of operations:
Brackets, Powers, Multiplication/Division, Addition/Subtraction.

Exercise 25A Order of operations and inequalities

1 Solve.

a $3 \times 4 + 7 =$ **b** $8 + 2 \times 4 =$ **c** $12 \div 3 + 4 =$ **d** $10 - 8 \div 2 =$

e $7 + 2 - 3 =$ **f** $5 \times 4 - 8 =$ **g** $9 + 10 \div 5 =$ **h** $11 - 9 \div 1 =$

i $12 \div 1 - 6 =$ **j** $4 + 4 \times 4 =$ **k** $10 \div 2 + 8 =$ **l** $6 \times 3 - 5 =$

2 Solve.

Hint Remember to work out any brackets first.

a $3 \times (2 + 4) =$ **b** $12 \div (4 + 2) =$ **c** $(4 + 6) \div 5 =$

d $(10 - 6) + 5 =$ **e** $3 \times 9 \div 3 =$ **f** $5 + 4 \times 2 =$

g $(5 + 3) \div 2 =$ **h** $5 \div 1 \times 4 =$ **i** $(7 - 4) \times (1 + 4) =$

j $(7 + 5) \div (6 - 3) =$ **k** $(8 - 2) \div (2 + 1) =$ **l** $15 \div (15 - 12) =$

3 Copy each number sentence and insert brackets to make the statement correct, where necessary.

a $4 \times 5 - 1 = 16$ **b** $8 \div 2 + 4 = 8$ **c** $8 - 3 \times 4 = 20$ **d** $12 - 5 \times 2 = 2$

e $3 \times 3 + 2 = 15$ **f** $12 \div 2 + 1 = 4$ **g** $9 \times 6 \div 3 = 18$ **h** $20 - 8 + 5 = 7$

i $6 + 4 \div 2 = 5$ **j** $16 \div 4 \div 2 = 8$ **k** $20 \div 2 + 2 = 12$ **l** $5 \times 3 - 5 = 10$

4 Jo says that $8 - 3 \times 2$ is equal to 10.

Is she correct? Explain your answer.

5 Amanda worked out $3 + 4 \times 5$ and got the answer 35. Andrew worked out $3 + 4 \times 5$ and got the answer 23.

Explain why they got different answers. Who is correct?

6 Here is a list of numbers, some signs and one pair of brackets.

2   5   6   42   +   ×   =   (   )

Use *all* of them to make a correct calculation.

7 Copy these statements. Insert either < or > to make each statement true.

a 5 ... 8   b 10 ... 3

c −6 ... 4   d −5 ... −3

8 For each inequality, write down the first three values for x, where x is a whole number.

a $x > 6$ b $x < 8$ c $x \leqslant 3$

d $x > 0$ e $x \geqslant 5$ f $x > -2$

Exercise 25B Units of measurement

1 Decide which metric unit you would most likely use to measure each of the following amounts.

a the height of your best friend

b the distance from school to your home

c the thickness of a CD

d the mass of your maths teacher

e the amount of water in a lake

f the mass of a slice of bread

g the length of a double-decker bus

h the mass of a kitten

i the temperature in Glasgow

j the distance between Dumfries and Inverness

2 Estimate the approximate metric length, mass or capacity of each of the following.

a the length and mass of this book

b the length of the road you live on

c the capacity of a bottle of wine

d the length, width and weight of a door

e the diameter of a £1 coin, and its weight

f the distance from your school to Edinburgh Castle

3 The distance from Stranraer to Aberdeen is shown on a website as 373 kilometres.

Why is this unit used instead of metres?

4 Sarah makes a living cleaning windows of houses. She has three ladders: a 2-metre, a 4-metre and a 6-metre ladder.

Which one should she use to clean the upper windows of a two-storey house? Give a reason for your answer.

26 Selecting and carrying out calculations

Exercise 26A Rounding, significant figures and estimating

1 Round each number to 1 decimal place (1 d.p.).

    a 3.73   **b** 8.69   **c** 5.34   **d** 18.75   **e** 0.423

    f 26.288   **g** 3.755   **h** 10.056   **i** 11.08   **j** 12.041

2 Round each number to 2 d.p.

    a 6.721   **b** 4.457   **c** 1.972   **d** 3.485   **e** 5.807

    f 2.564   **g** 21.799   **h** 12.985   **i** 2.302   **j** 5.555

3 Round each number to the nearest whole number.

    a 6.7   **b** 9.3   **c** 2.8   **d** 7.5   **e** 8.38

    f 2.82   **g** 2.18   **h** 1.55   **i** 5.252   **j** 3.999

4 Round each number to 1 significant figure (1 s.f.).

    a 46 313   **b** 57 123   **c** 30 569   **d** 94 558   **e** 85 299

    f 54.26   **g** 85.18   **h** 27.09   **i** 96.432   **j** 167.77

    k 0.5388   **l** 0.2823   **m** 0.005 84   **n** 0.047 85   **o** 0.000 876

    p 9.9   **q** 89.5   **r** 90.78   **s** 199   **t** 999.99

> **Hint**   Identify the required significant figure, and round correctly to this place value.

5 Write down the smallest and greatest possible numbers of people that live in these villages.

    a Ayeton   population 900 (to 1 s.f.)

    b Beeton   population 650 (to 2 s.f.)

    c Ceeton   population 1050 (to 3 s.f.)

6 When an answer is rounded to 3 significant figures, it is 6.14.

    Which of these could be the unrounded answer?

    6.140   6.143   6.148   6.15

7 For each calculation **a** to **d**:

    i estimate by rounding each number to 1 s.f.

    ii calculate the exact answer

    iii work out the difference between your estimate and the exact answer

    iv comment on the accuracy of your estimate.

        a 3.2×4.9 **b** 9.5×8.7 **c** 12.4×34.1 **d** 42.9×65.5

Exercise 26B Selecting and applying the correct operation to solve a problem 🗡

1. The train from Brighton to London takes 68 minutes.

   The train from London to Birmingham takes 85 minutes.

   a I travel from Brighton to London and then London to Birmingham.

   How long does my journey take altogether, if there is a 30-minute wait between trains in London?

   b How much longer is the train journey from London to Birmingham than from Brighton to London?

2. Michael is checking the addition of two numbers.
   His answer is 917.

   One of the numbers is 482.

   What should the other number be?

3. Jake, Tomas and Theo are footballers. The manager of their club has offered them a bonus of £5 for every goal they score.

   Jake scores 15 goals.

   Tomas scores 12 goals.

   Theo scores 20 goals.

   a How many goals do they score altogether?

   b How much does each footballer receive as a bonus?

4. a How many people could seven 55-seater coaches hold?

   b There are 288 students in eight forms in S5. Each form has the same number of students. How many students is this?

   c Adam buys seven postcards at 23p each. How much does he spend in pounds?

   d Phil jogs 7 miles every morning. How many days will it take him to cover a total distance of 441 miles?

   e In a supermarket, cans of cola are sold in packs of 6. If there are 750 cans on the shelf, how many packs are there?

   f Nails are packed in boxes of 144. How many nails are there in five boxes?

5. At 6:00 a.m. the temperature in Perth was −2 °C. By 2:00 p.m. it had risen by 14 °C.

   What was the temperature in Perth at 2:00 p.m.?

6. On the same day, the average temperature in Canada was −21 °C and the average temperature in Mexico was 41 °C.

   Calculate the difference in temperature between these two countries.

7. Greg the baker sells bread rolls in packs of 6 for £1.
   Dom the baker sells bread rolls in packs of 24 for £3.19.

   I have £5 to spend on bread rolls.

   How many more rolls can I buy from Greg than from Dom?

Exercise 26C Fractions, decimals and percentages

 1 Calculate these quantities.

a $\frac{1}{3}$ of £60   **b** $\frac{1}{4}$ of 84 cm

c $\frac{1}{7}$ of 224 g   **d** $\frac{1}{5}$ of 235 litres

e $\frac{2}{3}$ of £81   **f** $\frac{3}{4}$ of 68 mm

g $\frac{5}{8}$ of 192 kg   **h** $\frac{3}{5}$ of 585 ml

 2 Increase each amount by the given percentage.

a 340 g by 10%   **b** 64 m by 5%   **c** £41 by 20%

 3 Increase each amount by the given percentage.

a £80 by 5%   **b** 14 kg by 6%   **c** £42 by 3%

 4 Keith was on a salary of £34 200. He was given a pay rise of 4%.

What is his new salary?

 5 In 2004 the population of Lenzie was 14 200. By 2014, it had increased by 8%.

What was the population of Lenzie in 2014?

 6 A restaurant meal is advertised at £20. A service charge is added to all bills.

The bill for two people is shown.

Show that the service charge is 15%.

Romano's Bistro	
2 × Prawn Cocktail	4.60
1 × Chicken Risotto	6.60
1 × Sea Bass	8.50
1 × Fries	4.50
1 × Tiramisu	6.60
1 × Choc Fudge Cake	4.20
2 × Coffee	5.00
Service Charge	6.00
Total	46.00

 7 Mrs Denghali buys a new car from a garage for £8400. The garage owner tells her that her car will lose 24% of its value after one year.

What will be the value of the car after one year?

 8 In 2010 the population of a village was 2400. In 2014 the population had decreased by 12%.

What was the population of the village in 2014?

 9 A travel agent is offering a 15% discount on holidays.

How much will the advertised holiday now cost?

NEW YORK FOR A WEEK £540

 10 A shop increases all its prices by 10%. One month later it advertises 10% off all marked prices.

Are the goods cheaper, the same price or more expensive than before the increase? Show how you worked out your answer.

Exercise 26D Perimeter, area and volume

 1 Work out the perimeter of each shape.

a

5 cm
5 cm

b
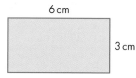
6 cm
3 cm

c
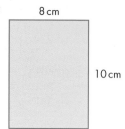
8 cm
10 cm

 2 For each rectangle, calculate:

i the area

ii the perimeter.

a

4 cm
4 cm

b
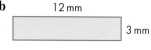
12 mm
3 mm

c
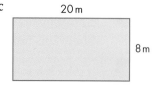
20 m
8 m

 3 Is it possible to draw a rectangle with a perimeter of 9 cm? Explain your answer.

 4 For each shape, calculate:

i the area

ii the perimeter.

a

3 cm
8 cm
3 cm
6 cm

b
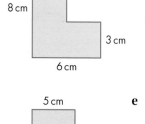
8 cm
2 cm
2 cm
6 cm
4 cm

c
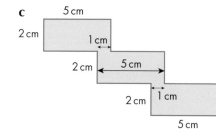
5 cm
2 cm
1 cm
2 cm
2 cm
5 cm
1 cm
2 cm
1 cm
5 cm

d
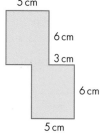
5 cm
6 cm
3 cm
6 cm
5 cm

e
10 cm
10 cm
10 cm
10 cm
10 cm
10 cm

5 Mr Jackson wants to fix laminate onto his kitchen worktop. The laminate comes in rolls that are 5 metres long and 0.5 metres wide.

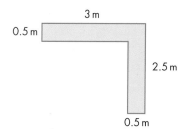

The diagram shows the plan view of the worktop.

a Work out the area of the worktop.

b Is one roll of laminate enough to cover the worktop?

Explain your answer.

6 For each triangle, work out:

i the perimeter

ii the area.

a

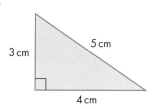

b

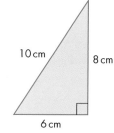

c

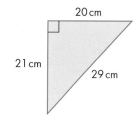

7 These compound shapes are made from rectangles and right-angled triangles. Work out the area of each shape.

a **b**

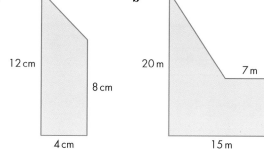

c

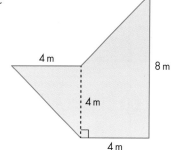

8 Work out the area of the shaded part of this shape.

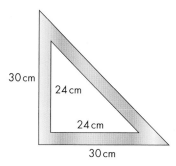

9 Work out the shaded area of each shape.

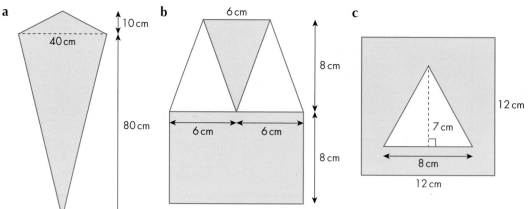

a 40 cm 10 cm 80 cm

b 6 cm 8 cm 8 cm 6 cm 6 cm

c 12 cm 8 cm 7 cm 12 cm

10 The diagrams show the dimensions of Mag's kitchen wall and the size of the square tiles she wants to use to tile the wall. They are not drawn to scale.

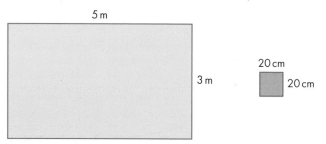

5 m 3 m

20 cm 20 cm

What is the minimum number of tiles Mag will need to cover the wall?

Hint Check that measurements are in the same units.

11 Which of these cuboids gives a volume closest to 3000 cm³? Justify your answer.

A 23 cm 12.5 cm 9 cm

B 10 cm 11.5 cm 22 cm

C 10.2 cm 10.2 cm 31 cm

Exercise 26E Speed, distance and time

> **Hint** The relationship between speed, time and distance can be expressed in three ways.
>
> $$\text{distance} = \text{speed} \times \text{time} \qquad \text{speed} = \frac{\text{distance}}{\text{time}} \qquad \text{time} = \frac{\text{distance}}{\text{speed}}$$
>
> $$D = ST \qquad\qquad S = \frac{D}{T} \qquad\qquad T = \frac{D}{S}$$
>
> Remember, when you calculate a time and get a decimal answer, do not mistake the decimal part for minutes. You must either:
> - leave the time as a decimal number and give the unit as hours, or
> - change the decimal part to minutes by multiplying it by 60 (1 hour = 60 minutes) and give the answer in hours and minutes.

 1 A cyclist travels a distance of 60 miles in 4 hours.

What was her average speed?

 2 How far along a motorway would you travel if you drove at an average speed of 60 mph for 3 hours?

 3 Mr Baylis drives from Jedburgh to Nairn in $4\frac{1}{2}$ hours. The distance is 207 miles.

What is his average speed?

 4 The distance from Leeds to Birmingham is 125 miles. The train I catch travels at an average speed of 50 mph.

If I catch the 11:30 a.m. train from Leeds, at what time should I expect to arrive in Birmingham?

 5 Hilary cycles 6 miles to work each day. She cycles the first 5 miles at an average speed of 15 mph and then the last mile in 10 minutes.

a How long does it take Hilary to get to work?

b What is her average speed for the whole journey?

 6 Marco drives home from work in 1 hour 15 minutes. He drives home at an average speed of 36 mph.

a Change 1 hour 15 minutes to decimal time in hours.

b How far is it from Marco's work to his home?

Exercise 26F Ratio and proportion

 1 The ratio of female to male members at a sports centre is 3:1. The total number of members of the centre is 400.

a How many members are female?

b What percentage of the members are male?

 2 A 20-metre length of cloth is cut into two pieces in the ratio 1:9.

How long is each piece?

3 Patrick and Jane share a box of sweets in the ratio of their ages. Patrick is 9 years old and Jane is 11 years old.

If there are 100 sweets in the box, how many does Patrick get?

 4 Marmalade is made from sugar and oranges in the ratio 3:5. A jar of marmalade contains 120 g of sugar.

a What is the mass of oranges in the jar?

b What is the total mass of the marmalade in the jar?

 5 Fred's blackcurrant juice is made from 4 parts blackcurrant and 1 part water. Jodie's blackcurrant juice is made from blackcurrant and water in the ratio 7:2.

Which juice contains the greater proportion of blackcurrant? Show how you worked out your answer.

 6 If four DVDs cost £3.20, what would 10 DVDs cost?

7 Dylan earns £18.60 in 3 hours.

How much will he earn in 8 hours?

8 A car uses 8 litres of petrol on a trip of 72 miles.

a How much would the same car use on a trip of 54 miles?

b What distance would the car travel on a full tank of 45 litres of petrol?

 9 Compare the prices of the products in each pair. State which offer is the better buy.

a

Mouthwash:

£1.50 for a bottle
£2.50 for a twin-pack

b

Deodorant:

£2.20 for 1
£4.45 for 2

c

Dusters:

49p for 5
95p for 10

d

Peas:

98p for 250 g
£2.75 for 750 g

  10 Compare the products in each pair. State which is the better buy. Explain your choice.

a Tomato ketchup: a medium bottle (200 g) for 55p, a large bottle (350 g) for 87p

b Milk chocolate: a small bar (125 g) for 77p, a large bar (200 g) for 92p

c Coffee: a large tin (750 g) for £11.95, a small tin (500 g) for £7.85

d Honey: a large jar (900 g) for £2.35, a small jar (225 g) for 65p

27 Reading measurements using a straightforward scale on an instrument and interpreting measurements to make decisions

Exercise 27A Reading measurements on scales

Hint	**Length**	10 mm = 1 cm
		1000 mm = 100 cm = 1 m
		1000 m = 1 km
	Weight	1000 g = 1 kg
	Capacity	1000 ml = 1 litre
	Volume	1000 litres = 1 m³

1 Copy and complete these statements.

a 155 cm = _____ m   b 95 mm = _____ cm   c 780 mm = _____ m

d 3100 m = _____ km   e 310 cm = _____ m   f 3050 mm = _____ m

g 156 mm = _____ cm   h 2180 m = _____ km   i 1070 mm = _____ m

j 1324 cm = _____ m   k 175 m = _____ km   l 83 mm = _____ m

m 5120 m = _____ km   n 8150 g = _____ kg   o 1360 ml = _____ l

2 Copy and complete these statements.

a 120 g = _____ kg   b 150 ml = _____ l   c 7500 ml = _____ l

d 3800 g = _____ kg   e 15 ml = _____ l   f 8.2 m = _____ cm

g 71 km = _____ m   h 8.6 m = _____ mm   i 15.6 cm = _____ mm

j 0.83 m = _____ cm   k 5.15 km = _____ m   l 1.85 cm = _____ mm

3 Read the lengths shown on this ruler.

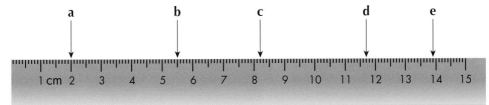

4 Jessie wants to replace the shelves in her living room. She measures the shelves she currently has fitted.

The first shelf's measurement is indicated on the ruler shown below.

Shelf 1

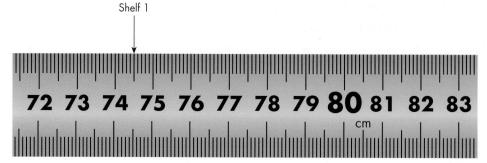

The second shelf's measurement is indicated on this ruler.

Shelf 2

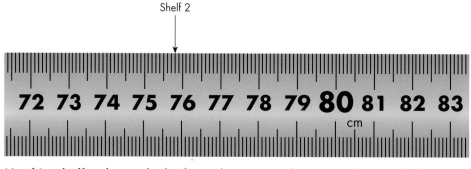

Her friend offers her a plank of wood 1.5 metres long to make into shelves.

Does she have enough wood for her new shelves? Give a reason for your answer.

5 Read these weighing scales and write the readings.

a

b

c

d

27 Reading measurements using a straightforward scale on an instrument
and interpreting measurements to make decisions

6 Mike is going on holiday. He can travel with a maximum luggage weight of 20 kg. He weighs his suitcase.

Mike notices that this is less than 20 kg. He weighs a pair of trainers on a different set of scales.

Can Mike add his trainers to the bag without going over the luggage limit? Give a reason for your answer.

7 How much liquid is in each jug?

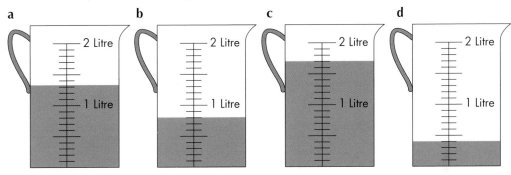

a   b   c   d

8 This measuring jug contains some juice.

William adds 700 ml of juice to the jug.

Will this be enough to fill it up to the 2-litre mark? Give a reason for your answer.

9 Write down the temperature shown on each thermometer.

a   b

c   d

10 The temperature on a night in Aviemore in January was recorded at 6 p.m. and again at midnight.

6 p.m.   midnight

By how much did the temperature drop between 6 p.m. and midnight?

28 Extracting and interpreting data from at least two different straightforward graphical forms

Exercise 28A Interpreting tables

1 Three TV companies offer different contracts at the same price.

Contract options	Company		
	A	B	C
Number of channels	316	137	382
Cost of broadband per month	£7.50	£5.65	£15.50
Upfront cost	£12	£42	None

Jean is looking for a TV contract which will give her at least 200 channels, but with a maximum broadband cost of £10 and a maximum upfront cost of £20.

Which company's plan would be best for her? Explain why you chose this company.

2 Mr and Mrs Brown are going on holiday to France. They want to travel on the ferry from Dover to Calais.

They have three options: Saver ticket, Flexible ticket or Premier ticket.

The table shows the ticket prices.

Type of ticket	Saver	Flexible	Premier
Cost per person	£39	£80	£105
Flexible departure time?	No	Yes	Yes
Wi-fi	None	30 min for 1 device	Unlimited

Mr and Mrs Brown want to pay no more than £90 each. They also want their departure time to be flexible and to have access to wi-fi for at least part of their journey.

Which type of ticket should they choose? Give a reason for your answer.

Exercise 28B Interpreting different types of graphical representation

1 This pie chart shows pets owned by people who took part in a survey. Each person surveyed had one pet or no pets.

a What fraction of those surveyed owned a dog?

b Make a comparison between those who owned a pet and those who didn't own a pet.

c What fraction owned a hamster?

> Hint | Use a protractor to measure the angle.

d 200 people surveyed said they owned a cat. How many people in total were surveyed?

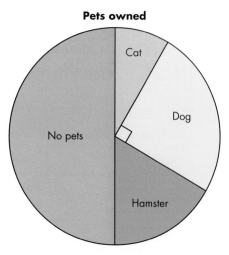

Pets owned

(Cat, Dog, Hamster, No pets)

2 A nationwide survey was carried out to research opinion on the friendliest region of Scotland.

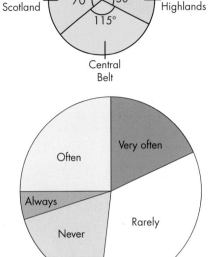

a What fraction of people picked at random from this survey answered Southern Highlands?

b If 7200 people were surveyed, how many people said:

   i Northern Highlands   ii Southern Highlands

   iii Central Belt   iv Southern Scotland?

3 Marion is writing a magazine article about healthy living. She asked a sample of people the question: 'How often do you consider your health when planning your diet?'

The pie chart shows the results of her survey.

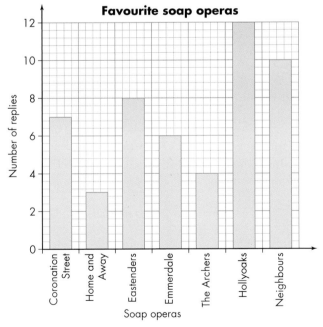

a What fraction of the sample responded *Often*?

b What response was given by about a third of the sample?

c Can you tell how many people there were in the sample? Give a reason for your answer.

d What other questions could Marion ask?

4 Linda asked a sample of people: 'What is your favourite soap opera?'

The bar chart shows their replies.

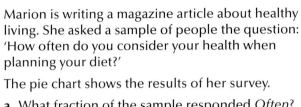

Favourite soap operas

a Which soap opera was chosen by six of the respondents?

b How many people were in Linda's sample?

c Linda collected the data from all her friends in S4 at school.

   Is this a good way to collect data? Give two reasons to support your answer.

5 The bar chart shows the results of 45 students in a mental maths test.

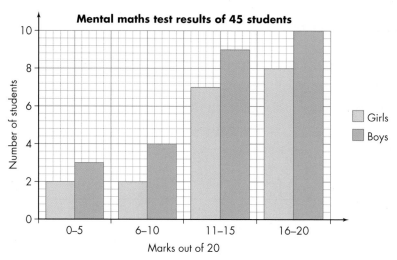

Mental maths test results of 45 students

Number of students (y-axis: 0 to 10)
Marks out of 20 (x-axis: 0–5, 6–10, 11–15, 16–20)

Girls
Boys

a How many boys scored between 11 and 15?

b Why is the bar chart misleading?

6 The bar chart shows the average annual rainfall of two major cities, one in England and one in Wales.

Elwyn says, 'There is more than three times as much rain in Wales as there is in England.'

Is Elwyn correct? Give reasons to explain your answer.

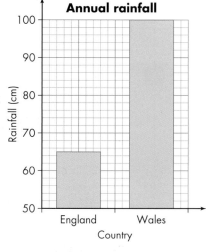

Annual rainfall

Rainfall (cm) (y-axis: 50 to 100)
Country (x-axis: England, Wales)

7 The line graph shows the monthly average exchange rate of £1 in Japanese yen over a six-month period.

a In which month was the exchange rate lowest? What was that value?

b By how much did the exchange rate fall between April and August?

c Between which two months did the exchange rate fall the most?

d Hamish changed £200 into yen during July. How many yen did he receive?

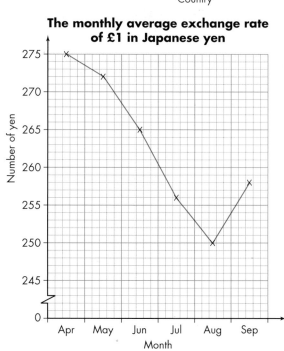

The monthly average exchange rate of £1 in Japanese yen

Number of yen (y-axis: 0, 245 to 275)
Month (x-axis: Apr, May, Jun, Jul, Aug, Sep)

28 Extracting and interpreting data from at least two different straightforward graphical forms

1 The man in the pictures is 1.8 m tall.

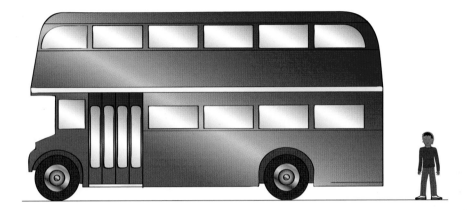

Use this to estimate the height of:

a a door in a house

b a double-decker bus.

2 The car in both pictures is 2.4 m long.

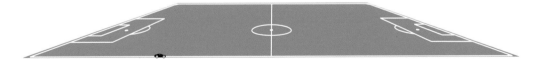

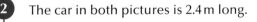

Use this to estimate the length of:

a an articulated lorry

b a football pitch.

3 The diagram shows the floor plan of a kitchen. The scale is 1 cm to 30 cm.

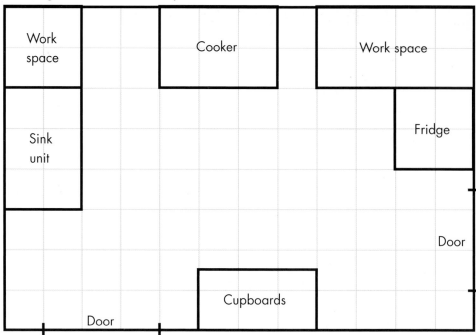

a State the actual dimensions of:

  i the sink unit   **ii** the cooker   **iii** the fridge   **iv** the cupboards.

b Calculate the total area of work space in the actual kitchen.

c Can a dishwasher measuring 65 cm wide by 55 cm deep fit into the space between the workspace (top left) and the cooker? Give a reason for your answer.

4 The sketch shows a ladder leaning against a wall.

The bottom of the ladder is 1 m away from the wall and it reaches 4 m up the wall.

a Make a scale drawing to show the position of the ladder. Use a scale of 4 cm to 1 m.

b Use your scale drawing to work out the actual length of the ladder.

c Can the ladder be used for a window cleaner to clean a window 6 m above the ground? Give a reason for your answer.

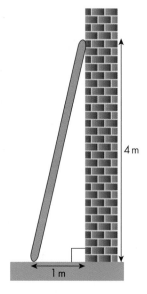

29 Making and explaining decisions based on the interpretation of data

Exercise 29A Drawing and interpreting graphs based on given data

1 The table shows the numbers of cars sold at a garage in one week.

Day of week	Monday	Tuesday	Wednesday	Thursday	Friday
Number of cars sold	9	10	11	10	12

Sam uses this information to draw a bar chart.

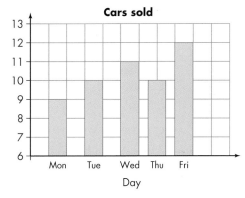

Cars sold

a A customer looks at the bar chart and says:
'Twice as many cars were sold on Friday as
on Monday.'

Is the customer correct? Justify your answer.

b Another customer says: 'This bar chart has
not been drawn properly.'

Write down **two** mistakes that Sam
has made.

2 Callum opened a new tea shop and was
interested in how trade was picking up over the first few weeks. The table shows the
number of teas sold in these weeks.

Week	1	2	3	4	5
Teas sold	67	82	100	114	124

a Draw a line graph for this data.

b From your graph, estimate the number of teas Callum can hope to see in week 6.

c Give a possible reason for the increase in the number of teas sold.

3 A kitten is weighed at the end of each week, for 5 weeks after it is born.

Week	1	2	3	4	5
Mass (g)	420	480	530	560	580

a Draw a line graph for this data.

b From your line graph, estimate how much the kitten would weigh after 8 weeks.

c Why might this not be a sensible estimate?

4 When plotting a graph to show the winter midday temperatures in Mexico, Pete
decided to start his graph at the temperature 10 °C.

Explain why he might have made this decision.

5 The table shows the estimated number of train passengers in a country at 5-yearly intervals.

Year	1970	1975	1980	1985	1990	1995	2000	2005
Passengers (thousands)	210	310	450	570	590	650	690	770

  a Draw a line graph to show this data.

  b From your graph, estimate the number of passengers in 2010.

  c In which 5-year period did the number of train passengers increase the most?

  d Comment on the trend in the numbers of train passengers. Suggest one possible reason to explain this trend.

6 The table shows the number of visitors to a museum from January to June.

Month	January	February	March	April	May	June
Number of visitors	14 800	14 200	14 700	15 300	15 500	16 800

  a Draw a line graph for this data.

  b Use your graph to estimate the likely number of visitors in July.

  c Between which two months did visitor numbers increase the most?

  d Explain this trend.

  e Is it possible to use this data to predict the number of visitors in November? Give a reason for your answer.

7 Amelie and Fran have tickets to see their favourite band in Glasgow.

They research the cost of a room in a hotel in Glasgow and the length of time it will take them to walk to the venue from each hotel. Their results are shown in the table below.

Time to walk (minutes)	1	7	3	9	16	12
Cost of room (£)	162	139	156	130	118	124

  a Draw a scattergraph to represent this information.

  b Write down the type of correlation shown by your scattergraph.

  c Draw a best-fitting line on your scattergraph.

  d Amelie and Fran decide they can afford to pay £150 for the room.

  Use your best-fitting line to estimate the length of time it will take them to get to the venue if they paid this amount.

8 Music pupils attending piano and guitar lessons sat exams at the end of the year. The table shows their results.

Instrument	Result					Total number of pupils
	Pass with excellence	Pass with distinction	Pass with merit	Pass	Fail	
Piano	208	888	1032	696	56	2880
Guitar	240	351	291	108	90	1080

  a Represent the data for each of the instruments in a separate pie chart.

  b In which exam (piano or guitar) did pupils perform better overall? Give a reason for your answer.

30 Making and explaining decisions based on probability

Example

Hannah buys 4 tickets for the school raffle. There were 125 tickets sold altogether.

Jonny buys 6 tickets for the youth group raffle. There were 170 tickets sold altogether.

Who has the better chance of winning a raffle, Hannah or Jonny? Give a reason for your answer.

$P(\text{Hannah wins}) = \dfrac{4}{125}$ •————— (Set up the probability in the same way as in Chapter 15.)

$= 0.032$ •————— (Convert to a decimal fraction (numerator ÷ denominator) to compare probabilities.)

$P(\text{Jonny wins}) = \dfrac{6}{170} = 0.035$ (3 d.p.)

Jonny has a better chance of winning, as 0.035 (3 d.p.) > 0.032.

Exercise 30A Making decisions using probability

1. Which is more likely: rolling a 5 or 6 on a regular fair dice or tossing a head on a fair coin?

2. Geeton football club have won 5 of the last 12 matches played.
   Kayton football club have won 7 of the last 16 matches.

   Based on past performances, who has the greater chance of winning their matches this weekend? Give a reason for your answer.

3. Kevin and Kendra are both athletes. They both race in 110–metre hurdles races.

   Kevin has come first in 3 of the last 20 races.
   Kendra has come first in 5 of the last 23 races.

   Based on past performances, who has the greater chance of winning their race on Saturday? Give a reason for your answer.

4. Do you have a greater chance of selecting the letter 'I' from the word 'MISSISSIPPI' or from the word 'ABILITY'? Give a reason for your answer.

5. Which is more likely: rolling two 6s on two regular fair dice or selecting the king of spades in a regular pack of cards? Give a reason for your answer.

6. In a 'square' game at the fairground, a player rolls a coin onto a grid with coloured squares. The player wins a prize if the coin lands on a blue square. There are 4 blue squares out of 36.

   In a 'triangle' game at the fairground, a player rolls a coin onto a grid with coloured squares. The player wins a prize if the coin lands on a red triangle. There are 3 red triangles out of 22.

   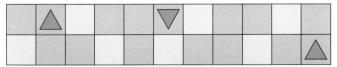

   Which do you have the better chance of winning, the square or the triangle game? Give a reason for your answer.

7 One bag contains 14 blue, 11 red and 5 green counters.
A second bag contains 6 blue, 6 red and 3 green counters.

From which bag do you have the greater chance of picking a green counter? Give a reason for your answer.

8 A group of S6 pupils sat their driving theory test. 12 out of 28 passed first time.
A smaller group of S6 pupils sat their driving practical test. 11 out of 24 passed first time.

Which test did these pupils have a greater chance of passing first time: the theory or the practical test?

9 Annabel travels to London regularly by train. Her train has been delayed 5 times in the last 45 journeys.
Annabel also travels to Paris by aeroplane. Her flight has been delayed twice in the last 25 journeys.

Is she more likely to be delayed on her train to London or her flight to Paris? Give a reason for your answer.

10 In the game 'Windup' you are dealt a card from a normal pack of playing cards. You win if you are dealt a 7 or 8.
In the game 'Upwind' you have to pick a letter randomly from the word 'PROBABILITY'. You win if you pick the letter 'P'.

Which do you have a greater chance of winning, 'Windup' or 'Upwind'? Give a reason for your answer.

11 Last year it rained on 7 days in February and on 8 days in March.

In which month is there the greater chance of rain this year? Give a reason for your answer.

12 In class S4A 3 pupils out of 24 are left-handed.
In class S4B 4 pupils out of 29 are left-handed.

If a pupil is picked randomly from each class, from which class are you more likely to select a left-handed pupil? Give a reason for your answer.

13 In which word do you have the greater chance of picking a vowel: 'COINS' or 'REGULAR'? Give a reason for your answer.